Water is Life

The late COLLEEN MCCULLOUGH AO (1937–2015) was one of Australia's best loved and most successful novelists. Born in Wellington, New South Wales, she first worked as a teacher, librarian and journalist in Sydney before studying medicine and neuroscience. From 1967 to 1976 she was a research associate in the Department of Neurology at Yale Medical School, and in the late 1970s settled on Norfolk Island, where she lived for the rest of her life.

Colleen McCullough's first novel, *Tim*, was published in 1974, followed by the internationally acclaimed *The Thorn Birds* in 1977. Between 1974 and 2010 she published twenty novels, including the seven-volume *Masters of Rome* series, and in 2010 wrote the libretto for a musical adaptation of her novel *Morgan's Run*, with music composed by Gavin Lockley, which premiered in 2011. She died in January 2015.

Also by Colleen McCullough

Tim
The Thorn Birds
An Indecent Obsession
A Creed for the Third Millennium
The Ladies of Missalonghi

The Masters of Rome series:
The First Man in Rome
The Grass Crown
Fortune's Favourites
Caesar's Women
Caesar: Let the Dice Fly
The October Horse
Antony and Cleopatra

The Song of Troy
Roden Cutler, VC (biography)
Morgan's Run
The Touch
Angel Puss
The Independence of Miss Mary Bennet
Life Without the Boring Bits (essays)
Bittersweet

The Carmine Delmonico series:
On, Off
Too Many Murders
Naked Cruelty
The Prodigal Son
Sins of the Flesh

here finally a better photo from someone who ate herself to death
without those two forwords the book would be impossible to publish.

This narrative by Colleen McCullough traces the career
of innovative Australian industrialist Michael Crouch AO
and was presented to him as a gift by the author.

Water is Life

*The story of water and the
Australian invention changing
the way the world boils water*

Colleen McCullough

With a foreword by Michael Jeffery

UNSW PRESS

A UNSW Press book

Published by
NewSouth Publishing
University of New South Wales Press Ltd
University of New South Wales
Sydney NSW 2052
AUSTRALIA
newsouthpublishing.com

First published 2016
'Water is Life' copyright © The Estate of Colleen McCullough 2016
'Foreword' copyright © Michael Jeffery 2016
'A word from Michael Crouch' copyright © Michael Crouch 2016
'Australian water projects' copyright © Michael Crouch 2016
Water is Life™ is trademarked.

10 9 8 7 6 5 4 3 2 1

This book is copyright. Apart from any fair dealing for the purpose of private study, research, criticism or review, as permitted under the Copyright Act, no part of this book may be reproduced by any process without written permission. Inquiries should be addressed to the publisher.

National Library of Australia
Cataloguing-in-Publication entry
Creator: McCullough, Colleen, 1937–2015, author.
Title: Water is life: The story of water and the Australian invention changing the way the world boils water / Colleen McCullough.
ISBN: 9781742235233 (hardback)
Notes: Includes index.
Subjects: Crouch, Michael, 1933–
    Zip Industries (Australia).
    Water – Boiling – Research – Australia.
    Hot-water heating – Research – Australia.
    Research, Industrial – Australia.
    Water resources development – Australia.
    Water – History.
Dewey Number: 609.2

Design Avril Makula
Cover design Alissa Dinallo
Cover images Aerial view of Darling Basin country in outback south-west Queensland during the wet season. Southern Lightscapes-Australia/Getty Images.
Printer Everbest, China

All reasonable efforts were taken to obtain permission to use copyright material reproduced in this book, but in some cases copyright could not be traced. The publisher welcomes information in this regard.

This book is printed on paper using fibre supplied from plantation or sustainably managed forests.

Contents

A word from Michael Crouch • ix

Foreword by Michael Jeffery • xiii

Water is Life • 1
by Colleen McCullough

Major Australian Water Projects • 107

Picture Credits • 129

Index • 131

A word from Michael Crouch

I am grateful to Colleen McCullough and her husband, Ric Robinson, for the many months she gave to the writing of this story. Colleen – a wonderful woman who suffered acute pain without complaint and whose love for Ric was paramount and obvious. I observed it when I first apprehensively took them both to dinner – it was at Eastbank on the Concourse leading to the Sydney Opera House. Her eyes sparkled at Ric. A remarkable, brave woman who fought pain for so many years, and an outstanding Ric committed to her welfare – they were a remarkable couple. I last saw Colleen when she visited me in hospital in 2014. A lively Colleen wheeled herself into the room communicating joy in a memorable way.

Earlier she had written to me enclosing the manuscript and said:

> The more I think about it, the less inclined I am to take any sort of payment: let *Water is Life* be a labour of love … Because I genuinely love you and admire you immensely, I wanted to produce something that would be a riveting read, first and foremost. Something neatly balanced between the story of water and the story of Michael Crouch, with a large dollop of Australia thrown in.

I had the good fortune to stay with Colleen and Ric at their home on Norfolk Island on a couple of occasions; their guest unit meticulous in anticipating any possible want. Importantly, I was able to spend a couple of hours looking through the many bays of Colleen's library. She had no computer and this whole story was written by her with thorough research, I assume, in her library and from her remarkable fountain of knowledge. A genius. Each page typed by her on her old typewriter.

The manuscript was completed in 2012 but I could only read it objectively when I had departed from Zip, as I did in December 2013.

The history of Zip is the story of people from many countries around the world who came to spend their working lives at Zip – all committed to building an Australian company that proved it is possible for an Australian manufacturer to export around the world and lead the world in both product and technology.

They were exciting decades and I would particularly like to pay tribute to all the members of the Zip teams in Australia, the United Kingdom and New Zealand, especially those who spent their time working inside factory walls and offices, some perhaps doing routine functions but performing their tasks with such commitment and dedication. It is these people whom my wife, Shanny, and my family, the beloved rocks of my life, salute with such admiration.

When Quadrant, a private equity company headed by Chris Hadley, purchased a controlling interest in the Group, the transaction, for me, was like divorcing one's family. The company, however, needed new, aggressive leadership. When I queried Chris as to whether the

A WORD FROM MICHAEL CROUCH

transaction would proceed, he said, typifying the great Australian spirit: 'You have my word'.

Since Colleen sent me the final manuscript, a number of people have assisted in bringing it to publication. I would particularly like to mention and thank Simon Best; Paul Brunton, for his research within the State Library of New South Wales; Linda Funnell, Colleen's long-time editor, for her meticulous work; Lee Kopcikas, my assistant, who typed many manuscripts; Gavin Lockley, who initiated *Water is Life* with Colleen; and Murray Pope, for his many readings of the manuscript. Importantly I wish to thank that great Australian Michael Jeffery, who has always so readily responded to my many requests and who contributed the Foreword. All these friends have spent hours assisting me in the most willing manner.

Above all, I hope this book communicates to you, the reader, the spirit of Australia, and that it reinforces the dire need to ensure, and highlight, the availability of an abundant supply of water across this nation. For this reason the book includes photographs showing some examples of how Australia has endeavoured to harvest and store water. Our attitudes, however, have yet to change. Much has yet to be done.

Michael Crouch AO
Sydney, July 2015

Foreword

The United Nations estimates that worldwide, 738 million people do not have access to clean potable water. While Australians are not in this situation, we are approaching physical water scarcity.

In her tribute to businessman, farmer, arts lover and noted philanthropist Michael Crouch, *Water is Life*, Colleen McCullough writes, 'Australia's political and bureaucratic arms of government do not take Australia's water situation seriously enough'.

However, Michael Crouch, the driving force behind Zip Industries for the last half century, 'knows that Australia must do more'. He knows too that ordinary Australians have little voice or, indeed, adequate awareness of our almost critical situation when it comes to water.

Water is life. Although there is enough water on the planet, it is unevenly distributed. Australia is an arid landmass where only its eastern coast and a small section of the south-western corner are capable of sustaining large human populations. Unlike the United States, Australia has very few river systems and our largest, the Murray-Darling, is much depleted. We rely, therefore, on dams that collect only 2 per cent of our rainfall and on our vast aquifer basins to provide irrigation water for our food production.

Colleen McCullough writes that Roman generals understood the basic need for access to clean potable water for the prevention of disease. In 1850 the British generals in Crimea did not, hence the massive loss of life to enteric diseases. Thousands of men died in World War I from appalling hygiene, resulting in cholera and typhoid. And it was not until the mid-twentieth century that people understood that boiling water kills the microbes that cause disease.

In 1962 much of Sydney's working population did not have healthy, hygienic hot water on tap. That year, Michael Crouch bought a small water heating company named Zip and later realised the value and potential of instant boiling water. For the next fifty years he improved the function and efficiency of Zip products. Zip has never deviated from its ecological precepts nor has it deviated from its dedication to the conservation of water in its efforts to reduce water usage and electrical power.

River systems and water have played a central role in sustaining human populations, but the United Nations reports that one-fifth of the world's population now live in areas of river water scarcity. Australians have the know-how and ingenuity to redress this situation but, to date, we don't seem to have the national will to protect our primary natural asset or make better use of the water we have.

FOREWORD

Michael Crouch's dream is to help solve Australia's age-old problem, the shortage of water. He has demonstrated one such way with technology and foresight in the development of Zip products.

He is determined to contribute to the protection of Australia's long-term sustainability in our continued access to clean potable water and agricultural, environmental and industrial water supplies.

Without water, human populations and ecosystems are critically threatened. Water is life. And for both the late Colleen McCullough and Michael Crouch this is a fundamental truth.

Major General the Honourable Michael Jeffery
AC AO (Mil) CVO MC (Retd)
Governor of Western Australia 1993–2000
Governor-General of Australia 2003–2008
Chairman of Soils for Life
Sydney, July 2015

Water is Life
by Colleen McCullough

WATER IS LIFE

The Pont du Gard, an aqueduct bridge in southern France, is part of the 50-kilometre Nîmes aqueduct, built in the first century AD by the Romans to carry water from a spring to their colony of Nemausus (Nîmes). The aqueduct carried an estimated 200 000 cubic metres of water a day to the fountains, baths and homes of the citizens of Nîmes, reflecting the value Romans placed on ample supplies of uncontaminated water – a priority not reflected in many later civilisations for centuries afterwards.

Water is life, a truth so manifest that no one stops to think about it. Let an organism or an environment cease to contain it, and death must ensue. Two simple statements, yet underlying them is a story as old as the universe and fully as complex. The story of water, this substance we take so much for granted, percolates in all directions, some surprising, some mysterious, always fascinating.

Taken as a percentage of global population, a fairly small number of us can call ourselves well educated: we read, we write, we arithmetically calculate efficiently enough to earn a living by using these skills. Among our privileged group is an even smaller group: those who are well educated in the sciences, who understand the structure of both a cell and a star, who appreciate the true impact of humanity upon our planet. They, the scientists, are schooled to observe with detachment, then form emotion-free conclusions that are properly valid. Emotions, as everybody knows, cloud judgement.

A general of armies in ancient Rome knew as a given that whenever he put his legions into a new camp, the troops had to draw their potable water from a stream having no settlement of any size along it, and that the latrine pits had to be dug as far as possible from the source of the potable water. If his troops were drinking well water, he knew separation of the two was of critical importance; he also understood how the lie of the land and its contours affected drainage, and also the properties of a water table. The Roman general had no idea why this obsession with the drinking–excretion cycle was so vital: he just knew it was because his training and the military manuals told him so. As indeed did his own experience. Unless he respected that cycle, disease would run through the ranks like wildfire; he had seen it happen as he rose up the chain of command and encountered a slipshod general.

The Crimean War of the 1850s is chiefly remembered today for the emergence of Florence Nightingale, who, with a band of helpers, nursed British soldiers in the Crimea as they died in thousands from the enteric triad – cholera, typhoid and typhus. What the Roman general at the time of Jesus Christ had known perfectly well, the British general of Queen Victoria's time did not. It was not the Russian enemy responsible for the near-defeat of the British, but the enteric triad of fevers, and they sprang directly out of appalling British camp hygiene. The British general of

> 💧 **The Roman general had no idea why this obsession with the drinking–excretion cycle was so vital: he just knew it was because his training and the military manuals told him so.** 💧

Queen Victoria's time had total contempt for the welfare of his soldiers, and cared not a farthing about the quality of their accommodation or their physical well-being. When the First World War rolled around in 1914, he still had the same attitude. Men died in thousands under machine gun fire, but they also died in thousands from the enteric triad in those filthy, mud-mired trenches.

Yet 'hygiene' was a uniformly known word by 1914.

Consider a single drop of water when it was first examined by the magnifying lens of a microscope several centuries ago.

If the drop of water came from a soupy village pond or the Thames River at London Bridge, it swarmed with what, in the seventeenth century, Anton van Leeuwenhoek called 'animalcules': invisible blobs of protoplasm that behaved not like plants but like animals, gobbling up or being gobbled up. Soon they were being called 'microbes'. But it was left to Louis Pasteur and others to make the inevitable conclusion, that these animalcules, or microbes, or germs, caused disease. Among the germs were the bacteria responsible for the enteric triad. Men like Joseph Lister then hypothesised that they could be killed by external agents known as disinfectants, or starved out of existence by being deprived of the nutrients they found in faecal and decaying matter; this was accomplished by extreme cleanliness – hygiene.

Few people today have any idea that nineteenth-century London was regularly ravaged by epidemics of cholera; thanks to typhoid fever in the British Royal Family, more became aware that typhoid existed. Yet how prevalent must lack of hygiene have been, that the Prince of Wales nearly died of typhoid? What chance did an East Ender have? Worst of all, there were medical and scientific men alive at that time who knew the answers; what they didn't have was the ear of Power and Money to improve sewerage, the water supply, and education about the need to be clean in the kitchen and clean in the lavatory. It would be halfway through the twentieth century before the general populace was educated and the great water/waste schemes finished.

However, one fact did spread far and wide early enough to make a mark in most households, and that was the discovery that boiling water killed microbes. Boiling an instrument for twenty minutes rendered it sterile, the word for free of all sorts of bacterial contamination. Boiling water renders it drinkable. And by the time 1900 arrived, most people knew that. Water *is* life. A drop of it reveals a world of life, plant and animal. But boil it, and life perishes unless it is a very strange and rare kind of life designed only to live in boiling water, and that need not concern us.

The first sterilising units in hospitals and clinics were based on boiling water, and it is still a technique available to anyone with a saucepan or kettle and a heating apparatus, be it a gas jet or an electric hotplate. Country living, depending on the quality of the water catchment, may demand that drinking water be boiled.

All this from a drop of pond water, a microscope, Anton van Leeuwenhoek, the enteric triad and hygiene!

New Zealand in 1946 was just emerging from a long period of austerity provoked first by a global economic depression, and secondly by a six-year-long world war. Housewives had done with very little in the way of improved domestic amenities. Hot water in any quantity was obtained by boiling water in the copper, a multi-gallon cauldron in the laundry or backyard; its reservoir was used to fill the bathtub to a tepid temperature, or to fill the wash tubs for laundering clothes. The general rule for the bathtub was one lot of water for the women, another lot for the men, and no lingering! There really was a ring around a 1946 bathtub! A large kettle boiled on the cooking stove provided the kitchen with hot water, which meant no dish was rinsed in hot water, and the roasting pan, washed last, required real elbow-grease.

However, there was the hot water urn, a device that dated back to the 1920s. A copper receptacle was wrapped in insulation, then put inside a steel skin; water was fed in at the bottom, where it heated, and a glass tube ran up its outside as a gauge; the moment the water boiled, it whistled.

During the Great Depression a New Zealand salesman, George Bigger, sold this hot water urn door-to-door for installation in factories or offices and was inspired to make improvements. A natural inventor, Bigger modified the kitchen/office unit to make it smaller, simpler – and patentable. Soon he was doing well enough to think that he should cross the Tasman to Australia and see how well he could do there. Setting himself up in Sydney, he did moderately well, making his heaters by night and selling them by day. Zip, as he called his enterprise, grew in size; by 1961 Bigger employed about a dozen people. Only one, however, was a salesman; Bigger was a better inventor than a businessman.

He concentrated on a hot water heater for the bathroom, the most neglected facility in most homes. The only improvement in its hot water supply since the backyard copper was the chip heater, a device that required a cold water tap to feed water through a funnel into a series of pipes in its skin; eventually heated water issued from a spout into the bathtub. The chip heater was fuelled by scrunched-up newspapers and tiny chips of wood, huffed and roared and spat, singed the eyebrows, burned the fingers, and was dangerous if the child delegated to light it decided to help the fire take hold with a cupful of kerosene.

The Zip bathroom heater was electric, had a shower arm for those who liked showers, and a tap arm for those who preferred a bath. It was safe and cheap to run, and it proved popular.

By 1959, George Bigger was ready to retire, and resolved to sell Zip Heaters and its patents. The terms of sale were expensive, but the young Australian businessman interested in buying Zip was not afraid of incurring debt. His family's business background served him in good stead, and he possessed a rare quality in any sort of man – foresight. So

> The chip heater was fuelled by scrunched-up newspapers and tiny chips of wood, huffed and roared and spat, singed the eyebrows, burned the fingers, and was dangerous if the child delegated to light it decided to help the fire take hold with a cup of kerosene.

negotiations began, and continued until March of 1962, when Michael Crouch and his family became the new proprietors of Zip. He was confident that his marketing techniques could cope with repayment, for he had already seen whereabouts Zip's marketing fell down: it worked more on a demand than a supply principle.

Any initial uneasiness sprang from the fact that George Bigger remained with the company to show the new owner the ropes; it was not a happy collaboration, if for no other reason than that the old broom hated the way the new broom swept. Once that problem was solved, Zip never really looked back; the other senior people in the firm from before Michael's time entered into his ideas and plans with wholehearted enthusiasm, and remained at Zip for years to come. No one was fired, though any increases in the workforce were thought about deeply before being implemented. Big economies were made in areas strange to the old Zip, like time-and-motion, proper division of labour, making sure round pegs were fitted into round holes and square ones into square holes.

In fact, the 1960s were sweet, Michael recollects. It was exhilarating to be a part of the new, growing Zip, founder-fellows of the family.

In 1964 the company moved to larger premises at Marrickville, financed completely by the ANZ Bank, smack-dab in the middle of a huge ring of Sydney's worst houses. Sales of bath heaters and shower heaters continued to grow, while low pressure water heaters also found buyers. This expansion was highly significant.

Michael set about manufacturing a supply for which better marketing would create a demand: this represented a radical change in the sales principle that did not, at first, sit well with his team. Leader foresight meant having to re-educate followers blinded by custom and tradition. As Michael saw Zip in 1964, its positive aspects lay in a performance superior to its competing lines, a superiority that rested upon patented innovations that competitors could not duplicate. On the negative side, Zip's marketing techniques were antiquated, haphazard and small-scale. What Michael had to do was market excellent products more shrewdly.

He asked for a new approach to selling. First had to come thorough investigation of what the potential market would bear; after that, production *capability* would be geared to supply it. His requests were greeted with horror, vehemently opposed, but he refused to back down. And as his sales staff began to see the wisdom behind his vision, all opposition gradually faded away. What had been opposition became enthusiastic co-operation and total support.

Owning a mind that always worked with logic and precision, Michael broke the components of an overall plan into steps that followed on each other chronologically, avoiding the chancy glitches that he *knew* had to end in a poorly devised plan of action. Those who worked with and for him grew more and more excited as he made them see that producing top-quality goods did not have to diminish profit. That if the marketing techniques were right, top-quality goods would increase profits.

In Zip, Michael Crouch found the ideal business he had looked for vainly in his twenties: the proprietor of a full process that he could lay his hands on all the way from the first glimmer of an idea to the last touch of gleaming chrome or enamel. That was top-quality, something to be proud of.

Which was fine and good, yet there were men who could congratulate themselves on an achievement without ever once stopping to wonder if there was a human being involved in the process. That can't be said of Michael Crouch, who is as proud of the people with whom he has worked over the decades as he is of any Zip product, no matter how many design awards it has won or how large the profits it has made. Perhaps here is not the place to discuss this side to Michael's character, but why not? It has often struck me over the course of our long conversations that everything he was, is, and will be, rotates around his membership and participation in the *gens humana* – the human family. If he sees his exquisite faucets and heating units as anything, I have come to the conclusion that to him, they are a human output, one of billions of items made by human hands out of human brains. Michael can never talk long without talking about people.

WATER IS LIFE

At the time of his taking over in 1962, Zip sales were mainly confined to rural New South Wales, and this continued. Good strong relations were established with the resellers in each town. There was a market in rural areas for bath and shower heaters, as the 1950s saw a rapid spread of electrification through New South Wales. It became possible for a rural cottager to install a 2.4-kilowatt apparatus in his bathroom, this being the minimum amount of electricity to power a Zip bath or shower heater.

By 1962, Zip was looking at those thousands upon thousands of Victorian-era workmen's homes that existed throughout Sydney's inner suburbs, including Marrickville, where the Zip factory sat. Michael initiated an exhaustive investigation of inner urban housing that yielded some glaring facts: that, in 1962, a high proportion of Sydney's working men lived in houses that were not conducive to either hygiene or comfort. A Zip heater over the kitchen sink and another in the bathroom would make living for these men's wives and families much easier – and healthier. It was every woman's wildest dream to have hot water laid on to the entire house, but economically that was impossible. What was possible was *almost* as good – hot water available at the kitchen sink and to the bathtub-shower, and for pennies a day in running costs as well as an affordable initial outlay.

By 1967, Zip had grown and expanded, but it was still a very small and dedicated team. Nothing deterred, it submitted an entry to the

The Zip factory in 1962.

The expanded factory required during the 1970s.

LEFT Early Zip kitchen water heater with sight gauge and ready whistle that blew when the water boiled.

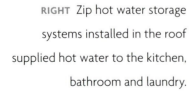

RIGHT Zip hot water storage systems installed in the roof supplied hot water to the kitchen, bathroom and laundry.

Hoover Award for Marketing, the only national marketing award given in Australia. In 1966, the winner had been Qantas.

The 1967 Hoover Award went to Zip. It created a sensation. Who had ever heard of this David among the business Goliaths? But from that moment on, Zip and Michael Crouch were never off the Australian business and marketing stage. Apart from a few minor investments, Michael had the good sense not to get bitten by the diversification bug and go into ventures unconnected to the heating and dispensing of water.

Which is not to say that improvements weren't made. The bath and shower heaters were redesigned to provide a better product for the market.

Zip realised that hot and cold water never mix. If the hot water could be drawn from the very bottom of the tank where the heating element is, and sucked up through a tube without mixing it with the cold water, then the system would work much better, as hot water would always be available from the top of the tank and would rarely run out of its hot water supply. It had another advantage – that the unit would only consume enough electricity to serve its actual needs. The heaters were remarkably efficient, and to cap everything, the device that drew the hot water off the bottom of the tank and kept it apart from the cold water was patentable.

By the 1970s the heaters were still popular and continued to sell, but Zip also started selling hot water systems it couldn't manufacture itself; these were always top-quality units purchased from the other leading hot water manufacturers, and offered people in country areas a comprehensive range of hot water systems. Time, however, to push the membrane again …

Coming into the 1970s – a decade of disappearing profits for Australian business – Michael made some resolutions. First and foremost, he wanted a healthy company that would attract investors if it ever went public; that meant a minimum profit of $100 000 per annum to qualify for listing on the Sydney Stock Exchange. After that, he wanted to produce a unique item or items that really worked with a minimum of trouble and servicing. He wanted to increase the size of his market. And, last but by no means least, he wanted to continue manufacturing his item or items within the Commonwealth of Australia. At that time Australia was a self-contained manufacturing nation, so the last resolution was feasible and achievable.

Everything was broken down into steps, even the nature of Zip's market. Nothing less than first place in any venue would satisfy Michael and his team, be it quasi-governmental outlets like county council showrooms and energy suppliers, wholesale appliance vendors, retail vendors from hardware stores to gadget shops, building contractors, building sub-contractors, even architects. Each was tackled in turn by the Zip sales staff, which led to the discovery that appliance manufacturers tended to forget the appliance once it had been sold on. So Zip went into the business of assuring its immediate market that a Zip product came with a formidable post-sales servicing record. From a converted bus that toured electrical appliance showrooms state-wide, to smartly grey-coated repairmen, Zip delivered on servicing, too.

Michael Crouch conducting a visitor tour of the Zip factory in the 1970s.

The response was immediate and compelling. Zip never looked back. At the heart of its enviable record, of course, was an appliance that delivered on its promises, well made, properly engineered, visually pleasing, as economical in its occupation of space as it was in purchase price.

All this from clean, hot water. Principles as old as the universe came together and were harnessed to improve the health and comfort of the Family of Man. Most busy executives don't pause to reflect on the more abstruse facets of their enterprise, but Michael Crouch always has. That brilliant, sparkling mind ranges far and wide, on, among many other things, the magic of water. Where does it come from, this substance that brings life?

ABOVE AND RIGHT Zip hot water storage cylinder manufacture in the 1970s.

Sydney, Australia, now a city of more than 4.5 million people, home base to Zip since 1947.

WATER IS LIFE

When the universe erupted out of nothing, it consisted of a sub-atomic plasma too hot for the human mind to comprehend. It was not until the birth convulsion lost some of its heat that the primordium quietened enough to form the first atoms, all hydrogen, the simplest of the ninety-two naturally occurring elements from which our universe is fashioned.

Each atom of hydrogen was a world in itself, having just one positively charged proton as a nucleus, with just one negatively charged electron zooming madly around it so fast that it masqueraded as a shell. The force of the Creation Moment blew this new matter, hydrogen atoms, farther and farther afield from the site of Ground Zero. As the hydrogen cooled, the galaxies and their stars were born, furnaces that caused the migration of multiple protons into the atomic nucleus, while an identical number of electrons zoomed madly around outside. Each additional proton–electron pair changed the nature of the atom by turning it into a new element. So helium followed hydrogen, with two, then lithium, beryllium, boron, each element heavier than the one before it. Hydrogen, however, always has reigned supreme. Some elements like to link up with others to form compounds called molecules.

Oxygen, in group 16 on the Periodic Table of the Elements, is an elderly element that has an affinity for linking one atom of itself to two hydrogen atoms to form a molecule of a compound we know as *water*. It has a freezing point and a boiling point, so can exist in any of three states – gas, liquid or solid – steam, water, ice. It is highly stable, which in chemical terms means it doesn't disintegrate under ordinary universal conditions, and its freezing to boiling point became the Centigrade scale for measuring temperature; scientists know that there is a cut-off temperature (far below the freezing temperature of water) for a substance's degree of coldness, and a temperature (far above the boiling point of water) at which the bonds holding molecules together fail to hold. Water can be written as a word, *water*, or as a chemical symbol H-O-H or H_2O. It is very common stuff.

Very common, that is, on Earth. Until relatively recently, no one dreamed that water might exist elsewhere.

Occasionally the Gods or a God or God sent the people of Earth an omen called *stella crinita* by the Romans and a comet by the English. Visually it is one of the most exquisite sights in the natural world: eerily silent, its gauzy tail streaming behind its glowing white ball of a head, it sits in the night sky, a celestial visitation. A portent, a precursor of doom.

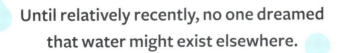

Until relatively recently, no one dreamed that water might exist elsewhere.

WATER IS LIFE

Comet McNaught, or the 'Great Comet of 2007', was the brightest to appear in the sky for forty years. It blazed across the sky of the Southern Hemisphere, visible to the naked eye even in daylight.

One comet marked the death of Julius Caesar. Another comet marked William the Conqueror's invasion of Britain. But when Edmund Halley predicted that *his* comet would return every seventy-odd years, divine intervention seemed less likely. Then came the huge shock of the 'dirty snowball' hypothesis: a comet was nothing more than a huge lump of dirty ice – frozen water from somewhere out in space.

Nowadays we know that comets are town-sized chunks of ice adulterated by many different elemental and molecular impurities. Over 4 billion years ago, our Sun was born; around it spun a richly gaseous sea called the Sun's proto-planetary disc. And one of the most enduring constituents of the disc was water. As the planets formed out of this elemental soup, things shook down in a logical physical way. Closest to the Sun came the rocky planets, then the gas giants, and finally, out in the reaches of the solar system, so cold the mere imagining of them is terrifying, ice planets and gargantuan icebergs.

Beyond the outer planet Neptune is a rotating expanse of town-sized chunks of ice called the Kuiper Belt. Hundreds of thousands of them all told, but that's chickenfeed. Just beyond the icy dwarf planet Pluto is an area bigger than the entire solar system in extent, but filled with a trillion vast icebergs and ice bodies. This is the Oort Cloud. Oort bergs don't move around the Sun like Kuiper bergs, they're too far from it for that, so they simply drift. This may give rise to a picture inside the mind of velvety blackness liberally sprinkled with massive isles of glittering white ice. The truth is blackness, nothing but blackness. The ice is so dirty it reflects none of the tiny light available so far out.

Halley's Comet, last seen in 1986, will reach its farthest point from the Sun on 9 December 2023, at a distance of about 5.3 billion kilometres, and will then begin to fall back toward the Sun. It will be best seen from Earth in late July 2061 when it will sweep within 80.4 million kilometres.

What we see in a comet sprawled gloriously across Earth's night sky is purely the result of the Sun, which is a star, with all a star's colossal power. Most comets originate in the Kuiper Belt, and it is a rare phenomenon for one of those town-sized chunks suddenly to veer wildly off-course and plummet inward to the Sun. As it embarks on this mostly suicidal fall, its substance is torn and ripped away by the buffeting solar winds to form the glowing tail. Its water we on Earth never feel, flung into the cataclysmic turmoil of inner space, whereas the comet's carbon, silicon, iron and rocky impurities burn up in Earth's atmosphere as meteor showers.

The 'dirty snowball' hypothesis said loud and clear that water did not originate solely on Earth, and that immense amounts of water Earth possesses originated in the relative emptiness of space. Since the days of space probes and physical analysis of extra-terrestrial objects began, knowledge about comets has mushroomed; scientists now understand, for instance, that the composition of a Kuiper Belt comet differs from that of an Oort Cloud comet. Just how many flavours does water come in, out there?

All of which is unimportant compared to the conclusion that water did not originate on Earth. If the two iceberg regions, Oort and Kuiper, are anything to go by, wherever there are stars with solar systems there will be water. As to whether there are any Earth-like planets or any life present – that is a completely different question. Though water says life is possible anywhere. Why? Water *is* life.

The smaller, rocky planets that formed around our Sun – Mercury, Venus, Earth and Mars – were molten bodies that gradually cooled and grew rocky crusts, albeit each after its own fashion. The proto-planetary disc's water was roiling among the other gases as each planet used its new gravity to capture a gaseous atmosphere. On Earth and Mars, some of the atmospheric steam condensed and fell onto the crust as rain. Acid rain at first. Venus still rains acid!

On Earth the acid content of rain diminished fast, for it rained a colossal ocean: millions of years of relentless rain. While the molten interior of Earth, not very far below, gently warmed the water, busy gathering to itself a host of different salts dissolved out of the crust – chemical building blocks. This tepid, nourishing bath provoked the elements oxygen, nitrogen, carbon, phosphorus, sulphur and some others to assemble into increasingly complex clusters electrically bound together in giant molecules, with water as the universal solvent and carbon as the most important template. Without water, none of it was possible.

These giant molecules, however, were not *alive* – merely large and complexly constructed. Life is a difficult state of being to define, but it involves function and change within a thing, be that a single molecule or an organism made up of billions of molecules. A molecule that does not grow or change or do work after it is fully formed is not alive. And again, water is the universal solvent, the substance which enables life to occur.

Not-to-scale visualisation of the relative position of our planet to the Kuiper Belt.

WATER IS LIFE

The 1960s saw Michael Crouch construct his human team, the men and women who, he knew, must believe in their product as ardently as he did himself. To understand the concept of teamwork is the benchmark of a wise and canny businessman, and Michael always believed in teamwork. In a way, he built a giant molecule out of individual units by keeping every member of his team, his staff, in the company loop; he made them feel that what they produced was of the best quality, and that their input, no matter how humble, was personally prized by him. Zip began to develop and cherish a sense of company patriotism among all its staff, who were thrilled by Zip's victories in the marketplace.

Which means that Michael had gone that extra, vitally energising step so few companies achieve: he had endowed Zip with *life*, for it grew and changed within itself after it was fully formed. He had taken his example from the raw material at the source of his product – water – and entered the 1970s with a vibrant, healthy company comprised of an utterly loyal and patriotic team.

There had been one stumble on the 1960s road, an inexplicable decline in profits. When research revealed the reason, it lay in a temporary loss of sight of the company's marketing strategy, and was easily rectified. But the lesson was there to be learned, and was learned: adhere to the original goals.

The 1970s in Australia were extraordinary years. In 1972 the latest of a series of conservative prime ministers was beaten by a Labor Party just emerging from over twenty years in opposition. It was led by a highly individual radical, Gough Whitlam, who within three years was dismissed from office by the Governor-General. Whitlam had spent the country's funds and one of his ministers was discovered attempting to borrow money overseas through peculiar sources; he had cut tariffs, given money away to local councils, cancelled the Imperial honours list and introduced the Australian honours system.

The chaos of the Whitlam years took several more years to overcome, as a new conservative government found a different nation. Business conditions had become very turbulent, and inflation raged throughout the early to middle 1970s. By the late 1970s, huge wage increases (the result of that inflation) made running a small manufacturing business extremely difficult.

A new phrase was creeping into business usage: research and development. Michael Crouch heard it, and heeded it. He had never been tempted by diversification, but saw research and development in one's own line of goods as mandatory if sales were to continue rising and the company to thrive. What Zip offered to the market was, if anything, shrinking, so that it could better concentrate on a few top-quality products well protected by the patents that come out of research and development.

Marketing hot water systems would have to go, the competition made profitability risky. Zip must find completely new products.

Slowly, the penny dropped. If all Zip's competitors were concentrating on devices that supplied *hot* water throughout the house for washing and bathing, Zip would not. Zip would concentrate on having *boiling* water dispensed from a heater the moment you touched its tap – boiling water instantly available for tea, coffee and cooking. A small, passionate team working in manufacturing came up with this magical product, and Michael saw how effectively it could – and, with proper marketing, would – replace the slow grumbling of an office or factory urn. Every office and factory in the country was a potential market, and Zip had the product to supply that demand: the world's first compact instant boiling water heater, that delivered water from its tap at exactly the same temperature as an electric kettle.

When James Watt built his steam engine in the eighteenth century, it contained a special part called a condenser, and Watt protected his steam condenser by a rigid, all-encompassing series of patents. Having no capital, he went into partnership with the Birmingham manufacturer Matthew Boulton, and together they had an absolute monopoly on steam engines. As steam is hot and can be kept under considerable pressure, steam has huge energy; without the steam engine, a function of water, there would have been no industrial revolution. To have a patentable advantage was vital for a small firm. Hence Michael's vision, research and development, and that small, passionate team who eventually made the break-through.

A water condenser became the heart of the new Zip instant boiling water system. A tank to hold the water that would be thermostatically controlled within 1 or 2 degrees of boiling point; steam would be collected via a small tube into a little condenser mounted alongside the boiling water chamber – if only James Watt knew! The condenser, the twin tank technology, became the heart of the Zip instant boiling water heaters that were to be used across the world in the course of the next two or three decades. The Zip boiling water heater was enthusiastically received.

The units became smaller in size as the 1980s appeared, though they had always been manufactured with an eye to smart good looks and clean lines. Michael could never see why a product's form and figure should not be at least as flattering as its function was excellent. The Zip Miniboil, a unit the size of a standard kitchen cupboard door, was an instant hit.

Until 1986 the company had confined its activities to the homeland, Australia. But in 1986 Zip went international by appointing distributors in Singapore, Hong Kong and the United Kingdom. It exhibited its revolutionary instant boiling water heaters in exhibitions in Germany in 1988, and later on obtained distributors in Taiwan, Thailand and the Philippines. South Africa soon followed, where its products are today manufactured under licence; in 1991 it commenced its own distribution company in the United Kingdom.

The continents were tumbling …

Instant Boiling Water	CI/SfB 51/53 (53.5)
	UDC 696.48

Kitchen Boiling Water

Zip Miniboil, for instant boiling water: a convenience every kitchen needs

In any kitchen, the last thing you want to do is waste time. Which is exactly what a jug, kettle or percolator does for you. They take time to fill. Time to boil, Time to refill. And more time to clean. None of which need worry you again once you install a Zip.

You don't wait a moment for water to boil. Just touch the tap and out it pours, a cupful or a potful at a time.

You get better tasting tea or coffee too. Because the water is boiling, not just hot. And because it is perfectly clean.

Zip helps with hundreds of jobs in your kitchen, from boiling baby's bottle to cooking cakes, making gravy or doing the eggs or vegetables.

You save money because Zip is more efficient. You save work because it refills itself automatically every time you use it.

You save mess because it takes up no kitchen bench space.

You save on maintenance in the long run because there's no steam to ruin paint, ceilings or curtains. No matter where you make tea or coffee, a Zip Miniboil can do a lot more for you—faster and better than any other boiling water appliance you can name.

Check the benefits. Check the savings. And switch to Zip Instant Boiling Water.

Instant Boiling Water

Zip Miniboil saves you time, money and energy every day.

Boiling water, instantly.
No more waiting for a slow jug or kettle to boil. Touch the Zip tap and out pours boiling water the instant you want it.

Refills itself automatically.
No need to worry about refilling. Zip refills itself automatically whenever you use it.

Uses less energy.
Unlike a kettle, Miniboil is fully insulated so all the heat and steam is trapped inside. And Zip doesn't waste energy by boiling more water than you use.

Zip Heaters (Aust.) Pty. Ltd.
22-26 Myrtle Street,
Marrickville NSW 2204
Telephone (02) 559 4144
Cables: Zipheaters Sydney

Vic. 428 0627; Qld. 358 4122, 391 8199;
S.A. 212 3161, 297 8700; W.A. 446 5400;
Tas. 34 2666; A.C.T. 80 5070, 80 6844

WARRANTY
Zip Water Heaters are covered by the Zip 5-year Extended Warranty. Each heater comes complete with written warranty and installation instructions.

Your Zip Dealer:

Never runs cold.
Touch the tap and out pours water within one degree of boiling point. Zip never runs cold, always delivers boiling water.

Always flows clean.
Built-in filter, copper boiling chamber and copper pipes mean your boiling water always flows clean.

Simple to install.
Any plumber can install a Zip Miniboil anywhere there is a power point, and a cold water service.

Built to last.
Every part is built to withstand constant use. Instant boiling water when you want it, year after year.

Fully guaranteed.
Guaranteed against faulty materials and workmanship for a full twelve months.

A quality Zip product.
Manufactured and serviced by Zip Heaters (Aust.) Pty. Ltd., manufacturer of Australia's largest range of hot water systems and boiling water units.

Miniboil Specifications.
Capacity 15 cup/2.5 litre,
Recovery 1 cup per min.,
Measures 350 mm high,
240 mm wide, 145 mm deep.
240 Volts A.C. 1500 watts.

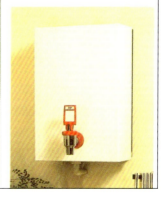

A two-page leaflet from the early 1980s introducing Zip Miniboil instant boiling water for home and office kitchens.

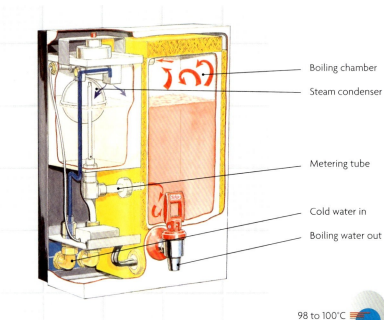

- Boiling chamber
- Steam condenser
- Metering tube
- Cold water in
- Boiling water out

Zip Miniboil provided an energy-efficient way to store water within 1 degree of boiling point, ready for instant use. Incoming cold water was sprayed into a condenser, where it was pre-heated by steam re-circulated from the boiling water storage chamber before being fed into the boiling water chamber for heating to boiling point.

- 98 to 100°C — Boiling water
- 65 to 80°C — Dishwashing water
- 38 to 44°C — Bathing water
- 10 to 20°C — Ambient water
- 5 to 10°C — Chilled water
- 0 to 4°C — Ice formation

WATER IS LIFE

The same principles are still used in electronically controlled, wall-mounted Zip Hydroboil instant boiling water appliances now used by millions of people around the world every day.

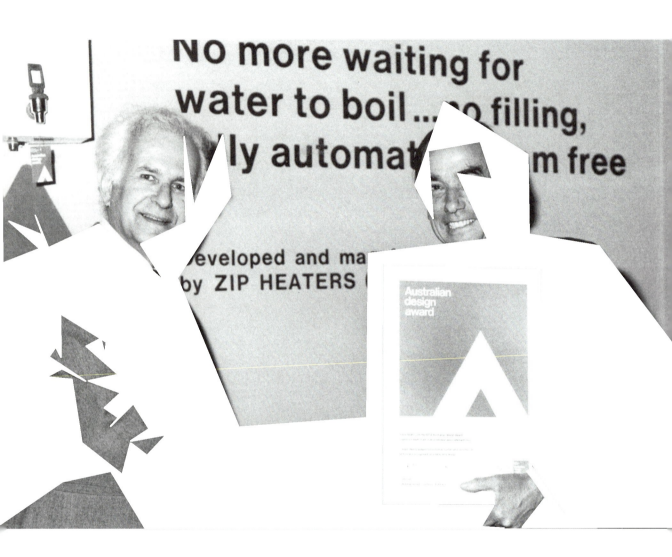

The launch of the Zip Miniboil instant boiling water heater for home and office kitchens was followed by the Australian Design Council presenting Zip with an Australian Design Award in 1984.

WATER IS LIFE

Within the last billion of Earth's 4-billion-year-old history, one gargantuan landmass formed: Pangaea. During the final third of those billion years, Pangaea split into two landmasses. One, called Laurasia, filled a very significant proportion of the Northern Hemisphere. The other, called Gondwanaland, tended northward, but did occupy some of the Southern Hemisphere. If today you look at a globe of the world, it's fairly easy to see that South America, most of Africa and Australia, together with Antarctica, fit together a little like a jigsaw puzzle. All of those modern landmasses were once a part of Gondwanaland.

Gondwanaland didn't fare nearly as well as Laurasia, whose sheer size endowed it with huge tectonic energy; it threw up new ranges of mountains, stole the Indian subcontinent from Gondwanaland, and built itself fresh, igneous defences against a much smaller ocean mass in the North Atlantic and the North Pacific. Nowadays there is a land hemisphere, the Northern, and a water hemisphere, the Southern; even Africa decided to tack itself onto Laurasia, thus forming the Mediterranean basin.

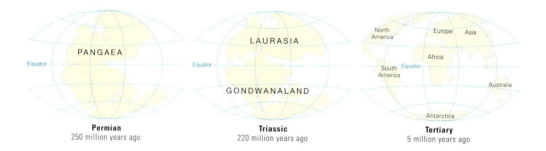

Earth's surface comprises some dozen tectonic plates that move in relation to each other at the rate of a few centimetres per year. Between about 250 million years and 220 million years ago the super-continent Pangaea split into two main landmasses, Laurasia and Gondwanaland, which then, about 5 million years ago, formed Earth's continents much as we know them today.

    Back now to the pools, ponds, lakes, seas and oceans that I left holding giant but unalive molecules. For reasons still largely a mystery, one kind of giant molecule succeeded in changing itself *after* it was formed. It did that by robbing the Sun of a tiny particle of energy, and used the energy to grow, to change – even, eventually, to reproduce facsimiles of itself. How did it do that last, reproduce itself? By originating a code within its matrix that contained the structure's pattern-card. It could duplicate itself at will or according to a cycle: it had found the secret that water had always held waiting for it: *life*.

    Life when it came went two ways: that of the plant, which retains the ability to rob its energy directly from the Sun; and that of the animal, which consumes the plants and other animals to obtain its energy. If you rob the Sun directly, you don't have to move in order to live your particular life; you can be rooted to the same spot provided that the water doesn't dry up. Whereas animals eat out their food supply, and must either move to find a new source of food, or move to cultivate the land to grow more food.

Laurasia ended in providing climates and environments ideal for the proliferation of life, of species. Whereas Gondwanaland stagnated. Though it owned some formidably vast jungles, jungles are not conducive to mammalian prosperity. What it lacked were temperate regions of equable climate and adequate rainfall, so its mammalian life didn't thrive at all by comparison with Laurasia. It had no bread-basket river systems like the Yellow, the Ganges, the Indus, the Nile, the Euphrates-Tigris, the Volga, the Danube, and Mississippi-Missouri-Ohio, to name a few.

Thus the plants and animals of Gondwanaland developed to suit a fairly water-poor environment. Southern oceans teemed with life, the super-continent itself did not. Archaeologically this is easy to deduce in the relative lack of prehistoric human habitations compared to Laurasia. Gondwanaland's temperate zones lacked permanent water sources, particularly from melting snows, as there were very few high mountain ranges.

WATER IS LIFE

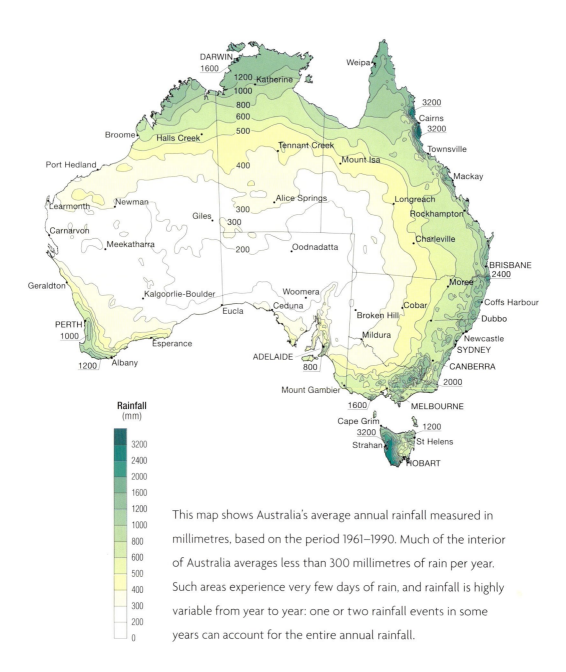

This map shows Australia's average annual rainfall measured in millimetres, based on the period 1961–1990. Much of the interior of Australia averages less than 300 millimetres of rain per year. Such areas experience very few days of rain, and rainfall is highly variable from year to year: one or two rainfall events in some years can account for the entire annual rainfall.

A large part of the inland registers an average annual rainfall that is so low that the land is too arid for crops.

WATER IS LIFE

The Australian chunk of Gondwanaland is both the globe's biggest island and smallest continent. To east, south and west of it lie abyssal oceans; to the north lie tropical, monsoonal islands that were once a land bridge. Parts of Australia have been covered by shallow seas from time to time. The most significant Australian mountain spine rises immediately behind the eastern coast, and in our own era is worn down to the stumps of a geological senescence. Alas for Australia, this location means that the rain falls to its east on a very narrow coastal strip. West of the Great Divide, as this mountain spine is called, the continent becomes largely a desert; what long rivers it holds are dry more often than running. Monsoon rains have given the northern fringe some verdure and two annual seasons: a four-month summer Wet, and an eight-month summer Dry. Spring, autumn and winter do not really exist.

Australia was never a land of milk and honey. Only its eastern coast and a small section of its south-western corner are capable of sustaining large human populations, and then mostly in cities, rather than scattered through bread-basket river systems; none exist naturally. For in all of this resides the same cardinal kernel, the substance that enables a plant or an animal to live: *water*. The substance in which life developed, the substance which permits life to continue. Without water, there is no life. Accept it as a mantra: *water is life*.

RIGHT Mountain regions with high average rainfall, such as the Yarra Ranges, have dense eucalypt forests.

The most popular surf beaches are located mainly in the high-rainfall areas of Australia's eastern and western coasts.

WATER IS LIFE

Zip too had a mantra, and a simple one: everything Zip made and marketed had to be of top quality. Problems and stumbling blocks were there to be solved, not accepted but hated, not swept under the carpet, not half-fixed. If it bore the word Zip, then it not only had to work; it had to work superbly well, top of its line.

In earlier years it wasn't too difficult to install a Zip hot water heater into bathrooms, but putting a boiling water unit into the kitchen often proved to be trickier. Where *did* one install an over-the-sink hot water unit? The space was almost inevitably filled up with windows and wall cupboards.

Returning to the Zip mantra about top quality, whatever was visible to the householder's eye had to look absolutely spiffy, a point of pride Michael Crouch and his Zip team refused to compromise. A Zip unit was immaculately finished, well-shaped and designed, and came in sufficient colours to find the right one for a particular kitchen's décor. In Miniboil and the new Zip Hydroboil, Zip had something top-class to offer the world, and the world accepted them with enthusiasm. In 1984, Miniboil won an Australian design award, the ultimate accolade for a product's excellence.

The story of Zip is replete with decent business ethics, and stands as proof that they work in the marketplace, despite the cut-throat tactics of much modern business and the drive to off-shore manufacturing in foreign locations where standards are not enforced so rigidly and extra cents can be made thereby. Things Zip has avoided, yet a healthier, more profitable business would be hard to find. Research and development germinated spontaneously from questions asked about potential markets, from wish-lists made by the sales force, from a chance remark made to a housewife. The big difference between Zip and other firms was that the executives of Zip, culminating in Michael Crouch, always listened and thought. Everything was grist to the Zip mill, and as the market changed in response to varying national conditions, the improvements and expansions kept coming because no one grew too complacent.

Rented premises became owned premises large enough to cope with increased demand, but always while Michael and Zip continued to adhere to Michael's two most cherished ideals. One was to regard

> **Research and development germinated spontaneously from questions asked about potential markets, from wish-lists made by the sales force, from a chance remark made to a housewife.**

every Zip employee as a part of the grand scheme, a vital member of the family, a person with a name and a life – no Zip employee was ever just a cipher. His second ideal was to make sure that every product, no matter whereabouts in the world it might be sold, must bear the declaration 'Made in Australia'. He implicitly believes that Australian manufacture is the very best.

The 1970s had been savage years, but the early 1980s and the late 1980s were even worse, particularly Mr Paul Keating's 'recession Australia had to have'. Little surprise, perhaps, that the Eerie Eighties saw Michael and Zip's team begin to look at water in different ways and, in particular, at one device that had changed hardly at all since the days of Queen Victoria: the tap, or faucet, or valve. Why not have a tap that could produce *boiling* water? Not warm, not hot, but physically boiling.

Kitchen space had been found: set Zip a problem, and Zip will find an answer. In this case, the answer was one of the most overlooked and apparently useless spaces in a house: the cupboard under the kitchen sink. Sink bottoms and mazes of pipes make it impossible to shelve; about the most a housewife does with its stygian gloom is hang her dish cloths to dry over the elbow bend in the drain pipe, and put her disinfectants and toxic bottles on its floor. When Zip first started installing kitchen sink hot water units, the modern kitchen with its chrome and Laminex didn't exist, but even after it did, the cupboard under the kitchen sink remained an ideal site for the bulk of a water heating unit.

Development of the Zip HydroTap took a number of years; by the end of the 1980s it still hadn't seen the light of day. In fact, it had its debut in

1996, at a time when Michael's integrity and acumen were known far and wide in his own country, and known abroad too. His personal eminence had grown to the point whereat the Prime Minister of Australia, John Howard, asked him to represent Australia on APEC (the Asia-Pacific Economic Cooperation) – but more of that anon.

Here is the place to talk about Zip HydroTap and the range that came from it, for Zip kept on improving its products. It never stood still. To cap its triumphs, the HydroTap gamut of taps won a number of Australian design awards.

Zip HydroTap is amazing. In one sleek, soaring, slender chrome pillar lie all the advantages of both instant boiling water plus chilled filtered water. One gets all the advantages of boiling water without waiting for a kettle to boil and all the benefits of chilled, bottled water without the eco-waste of bottles. The Zip HydroTap has a built-in water filter that removes chlorine and other contaminants from tap water.

The history of Zip is as steady as it is steadfast. Eager to put his hands on this manufacturing and marketing adventure as a young man at the beginning of the 1960s, Michael Crouch has always been wise enough to stick to products he and his team know, avoiding the snares inherent in diversification. The jungle is enormous, its contents nigh as varied as the number of stars in the night sky, but Michael still hews a well-known path through that jungle amid contents Zip is famous for. Not for him, to wander out of his beloved chosen sphere.

Another achievement for Zip in 2004: a boiling and chilled filtered water Zip HydroTap with lever control and a font kit allowing it to be installed remote from any sink.

From 2012 onwards, Zip HydroTap models included a range that, at the touch of a tap, gave not only boiling and chilled filtered water but also chilled filtered sparkling water.

Spoil yourself with a Zip HydroTap®

Your Zip HydroTap® brings you... boiling filtered water, instantly, chilled filtered drinking water, plus **sparkling** chilled water, too, filtered 25 times finer than ever.

 0.2 micron filtration

 Power-Pulse™
Cuts power costs five ways.

Zip HydroTap®
Boiling and chilled filtered water, instantly.
Sparkling, too!

 Scan to view video.
Call 1800 42 43 44 for a brochure.
Or visit www.zipindustries.com

WATER IS LIFE

Chilled filtered sparkling drinking water from the Zip HydroTap was a big hit with the capacity crowd attending the TEDx 'Ideas worth spreading' conference at the Sydney Opera House in May 2013. Free serve-yourself water stations were provided throughout the venue.

WATER IS LIFE

Consider, for a minute, those arch-wanderers, Lucy and her cousins, the very first human beings of a mere 8 million years ago. They roamed the land of their origin, Africa, but it, like Australia, was not theirs for the tilling. Far too unkind, too unforgiving. Then some descendants of Lucy's made the trek northward through the Nubian canyon of the Nile and emerged into Egypt, where, on either side of the great river, edible grasses grew and the food animals could be confined in flocks. Simultaneously, others among Lucy's descendants arrived in a land where two great rivers, the Euphrates and the Tigris, ran almost side by side through rich dark soil, and the same edible grasses grew.

Someone discovered that the seeds of eared grasses, if crushed, mixed with water and baked in the coals, made a tasty food: bread. Probably as an accidental outcome of slipshod housekeeping, yeasty animalcules contaminated the bread dough, which rose and, on being baked, turned into the staple food: leavened bread. Which came first – Egypt or Mesopotamia – is hotly debated, but both places were fed by bodies of perpetually replenished water that enabled people to live permanently on the same farms for generations, and gave them sufficient food to produce plenty of healthy children. Permanence of location gave rise to work cycles, and to their concomitant, leisure; fruitful ground for new inventions, ongoing improvements, and social ingenuity.

Ploughshares, swords and crowns translate as peasants, soldiers and the king. Society could be organised and stratified, thereby guaranteeing survival of the ever-multiplying group. All stemming from a system of rivers that gave perpetual water. In other parts of the globe bread-basket river systems behaved in the same way, though the edible grass might be rice in the tropics or rye in cold temperate zones. A handful of people could scrape a living nearly anywhere, but large populations needed the sustenance of flowing water that was present because of topographical features like mountain ranges capped with perpetual snow. Rivers and seas were the first roads; Man had the boat before he had the wheel. Water was more than merely life. Water enabled life to spread across the globe, including human life.

As seen in the example of the Roman general and his legions, the Romans revered water as life. It was they who understood the vital necessity of bringing sources of fresh water into urban areas to combat the diseases of squalor: they taught even the abjectly poor to revel in the sensation of cleanliness by building them free baths. Like the Greeks, they understood the mechanics of human waste disposal, sewered their cities and did not contaminate their ground water by burying their dead. Cremation was healthier.

When Christianity succeeded Imperial Rome, the significance of keeping both the water and the populace clean was lost. Even in eighteenth-century London, the sole source of water for a nobleman's marble palace was a 1000-litre water butt that collected rain from filthy roofs. Enteric fevers like cholera and typhoid raged, food poisoning was rife, and neither clothes nor persons were often washed. Hot water in the nobleman's marble palace was dragged in buckets all the way from the kitchen to the tub by the fire in the dressing room that saw the nobleman enjoy the luxury of a tepid, rather brown-watered bath.

Something Burke and Wills, Ludwig Leichhardt and the other early European explorers of Australia thought a lot about as they inched along, always dreaming of finding a Mississippi, or dipping their toes in a Great Lake. Instead, they found waterholes, dried-up stream beds, salt lakes, the racing torrent of a cloudburst gone in an hour.

Australia is an arid landmass, its water all but dwindling.

Zip has always been dedicated to the conservation of water and the conservation of the environment, though when Zip began, an ecological side to things wasn't mentioned.

Instead of wasting electricity and water overfilling kettles and urns, Zip always aimed at providing only enough boiling water for actual use through having a range of models to serve varying numbers of people. This saved consumer pennies, yes, but it also spared the environment the outcomes of wasted energy – the generation of electrical power and atmospheric pollution from that. Not significant in terms of one Zip

Miniboil, but add up the thousands and it does start to make a difference. Go into the hundreds of thousands, and …

Though the huge, ecological picture wasn't a part of Zip's marketing techniques in the early years, Zip salesmen could always point to those savings in water and power: Zip knew the ecological ground rules by instinct, arising out of its own mantra – good design, top quality Australian manufacture, and strict economy in consumption of energy.

Zip never deviated from its ecological precepts; and in these present days of general public awareness, can point proudly to its own record. Simply, as the eco- and energy-conscious 1990s rolled on into the Third Millennium, Zip never had to re-plan or re-tool to obey changing laws. It just went on in its perpetually eco-friendly way. Don't buy bottled water and throw away the bottles! Buy a HydroTap instead! Have a cup of tea or a glass of chilled water, both super-filtered!

Zip HydroTap makes a positive contribution to sparing the environment. No energy is wasted boiling more water than is needed. Having chilled drinking water on tap, and sparkling water too, reduces the need for bottled drinking water and soft drinks, thereby reducing the growing problem of discarded bottles and cans.

WATER IS LIFE

Thus far I have concentrated upon Zip and its genesis, its onward progress, but I have said nothing much about the man who not only stood at the Zip helm for fifty-two years, but was also its inspiration, its guiding light, its water: Michael Crouch.

What is he like today, as he whizzes around the world in pursuit of business or philanthropic ends at an age when his peers are lying on sun lounges or languishing in nursing homes? The laurels of victory sit firmly on his brow, he could relax; his hand has guided Zip through the Sweet Sixties, the Savage Seventies of disappearing profits, the Eerie Eighties and its fight to survive, the Nasty Nineties and their fight for profit, and the sound growth of the earliest Third Millennium. He's not a fool, so there have been holidays and vacations, rests when he needed them; this is not a man chained to his desk, nor ever was. More accurate to say, this is a man on the move, always on the move.

He's a neat, compact, wiry man who moves with a quick surety born of a confidence that is very well embedded. As a young man he would have been called handsome, with regular features atop good bones. His hair and eyes are dark, his skin ruddy in a way that says the Sun tends to burn him, so long exposure to the Sun has lent his complexion a hint of mahogany. He's articulate, eager, curious, and filled with strong opinions, but even a very short meeting tells you that here is someone about as up-front as people get. A straight man of unbendable ethics and principles.

I understand why Michael Crouch is so up-front: his life is just too busy to permit of anything else, too stuffed with incident and drama to bother with the tangled webs of deception. I think that long ago he discovered it was more comfortable and less time-consuming to be the same man on the surface as beneath.

What intrigues me most is how Michael escaped the taint of coldness, greed and self-absorption present in so many Australian businessmen, politicians and public service mandarins. The answer must lie in a combination of his basic character with his particular childhood. One astonishing quality stood out for me when I first met him: his innate humility. Yet another was his boundless and unquenchable energy. The clock must tick into old age, but rare men like Michael Crouch know how to take youth's advantages with it, so they never grow old. Fifty, tops …

In balancing nature against nurture, there is one inborn asset that puts a limit on what can be accomplished, in what areas. That is the brain a human being receives as part of the birth package. The most extraordinary men and women emerge as achievers from the most dismal of backgrounds, while the son and heir of an emperor is born a gullible fool unfit for office. And in between the two extremes comes a vast parade of people born with a top-quality, first-class brain. Were Michael Crouch not gifted with such a brain, Zip could not have happened, for out of the dreams, ideas, thoughts and aspirations of Michael's brain, Zip was formed, cared for, guided and encouraged. Its success is a measure of Michael's intelligence, a word that must contain wisdom and experience if it is to be properly defined. He reads, he travels, he observes; he takes what he sees, hears, touches, smells and tastes and submits them to the discipline of his thought processes, and the world is the richer for it.

Michael Crouch was born into easy circumstances in Sydney, Australia, during the Great Depression of the 1930s, but not into an atmosphere of waste. His father, a successful businessman, was a strict domestic economist who insisted the leftover roast be minced up into a shepherd's pie, and never left a light switched on for a second longer than it was absolutely needed. He saved petrol by coasting downhill when he drove, yet picked up people waiting for public transport because it appalled him to drive an empty car. When it came to his sons' education he didn't stint; Michael was sent to Cranbrook, an Anglican private boys' school in Sydney's eastern suburbs. There Michael received an excellent education, though unfortunately marred by periods of protracted illness that kept him at home. This meant Michael's school-based friendships were interrupted and gave him rather a lonely childhood, but it wasn't all bad, not with that restless intellect marooned inside a sick boy. He dabbled in things he fancied trying to pass the hours, from weaving to making crystal sets to playing the piano; he had a marked love for music. And everything is grist to the mill of a restless, top-quality mind, with one thing at its pinnacle – reading. Michael read books of just about every kind; he was an omnivorous reader.

It was always assumed that he would go into business; Michael assumed it too. What he wasn't *quite* so positive about was whether his line of business should be the same as his father's. The father ran an importing firm, and Michael's elder brother was already in it with their father. Michael's fate was assumed, and to prepare for it, he enrolled in Economics at university. Oh, but it was *boring*! With business set, as it were, within family concrete, his whole childhood had been a preparation for what, up at university, he often felt either a boring repetition, someone's airy-fairy theory, or plain untrue according to experience. Was it even possible to make a science out of an activity whose health seemed indicated by the height of women's hemlines?

Joining the family firm didn't work either; it was too confining, too much a re-enactment of childhood, the hierarchy unchanged. That enquiring, highly intelligent mind inside Michael's head was brimming

> **It was always assumed that he would go into business; Michael assumed it too. What he wasn't *quite* so positive about was whether his line of business should be the same as his father's.**

over with ideas – ideas that, he could see, could not be pursued in any but an environment he himself created and he himself stood at the head of. He even had a little capital of his own that could form the nucleus of investment capital, could he only find the right venture. As part of his quest he had joined Lodge Cranbrook early on, for Lodge Cranbrook was stuffed with businessmen; surely one of them would point him in the right direction while he was still young enough to seize new ideas and new projects eagerly. And that is exactly what happened. A friend in Lodge Cranbrook told Michael in 1959 about a New Zealander in the market to sell his little company and life's work to fund his retirement. The name of the company was Zip Heaters (Aust.), and its owner, George Bigger.

As said earlier, the purchase didn't happen overnight, and as this firm incorporated patents, as well as considerable good will, final closure didn't occur until March of 1962. What appealed most about Zip was its size. There would be no sitting isolated in an ivory-towered office for its proprietor; he would be called upon to participate in all aspects of company function, from marketing techniques to the factory floor – hands-on all the way! The range of products offered was small, yes, but they were very well designed and manufactured. The idea of using water as the end-product was fascinating. Most importantly of all, Michael saw that if he owned the business, he could make a bigger and better business out of it while scrupulously adhering to its already high standards of manufacture.

But the story of Zip is told; here is the place for Michael Crouch's personal story, including his involvement with the business as well as what one might call his private life. In Michael's case, it's all inextricably intertwined.

Which means other lives are closely connected to Michael's, lives that function as a part of the Zip team. It has expanded over the years since 1962, at present occupying spacious premises in Condell Park, Sydney. Zip Heaters (UK) was established in 1991, with its head office near Norwich, Norfolk, and a marketing office in London. The New Zealand subsidiary, Zenith Pty Ltd, came into being in 2006. Thus by 2012, Zip had become the Zip Industries Group. However, none of the original team spirit and enthusiasm had been lost, and this was especially so for those on the factory floor, people who, day after day, performed so many routine tasks to ensure the quality of a Zip product.

The Group enjoyed unprecedented growth in its recent history, empowering it to change the way the world boils water. By 2012 Zip was exporting to more than sixty countries, and Zip products are used by millions of people each day around the globe.

Michael was well into his thirties and still single when a happy fate intruded: he married his beloved Shanny, spiritually a country girl, passionate about the land and its cycles and activities. She was to prove a true life's partner who contributed her full share to a very long and

WATER IS LIFE

Michael and Shanny Crouch at home in the Hunter Valley.

satisfying union. Charlotte and Sarah, their daughters, are like Shanny – happiest in the country, each with the ability and warmth to light up a room as far as their loving parents are concerned. In George, their son, Michael received the greatest gift possible for a father: a son who follows Dad into the family firm wholeheartedly. He prepared himself for an eventual Zip executive post with a long and rigid apprenticeship that took in every aspect of Zip's output and interests, even to working in the meat industry, because the Crouch family has a large pastoral property, Waverley, in the upper Hunter Valley. No stone was left unturned.

By the time George considered himself ready for it, the appointment of Michael's son to executive management was a huge boost for the entire

Waverley Station, a major Angus cattle breeding property north of Gundy, near Scone, in the upper Hunter Valley, has been the country home of the Crouch family since 1991.

Zip team, from factory floor (where George had done time too) to his fellow executives. What Michael likes most about his son is their sharing of the same ideals, particularly the one that demands Zip products be made in Australia. As a man, George impresses his father with his quiet competence and the feeling he generates in those around him of solidarity and continuity.

Michael and Shanny's son, George, and daughters, Charlotte and Sarah, at Waverley Station.

Though business demands that there be a house in Sydney, *home* is the Angus beef property in the upper Hunter Valley. Shanny prefers rural living, so do the girls; it was Mum who taught the kids to ride, taught them the ways of the bush. Michael always speaks Shanny's name with enormous affection and respect, in that casual way betraying her constant presence in his thoughts, a part of him and his spirit.

And what a place to live, the Upper Hunter! Surrounded by the glorious forests and cliffy gorges of the Great Divide.

Though its course is short and it is subject to terrifying floods, the Hunter River is about as close to a natural bread-basket river system as Australia gets. The Hunter *flows*, its catchment a vast region of wet or dry forests dissected by gorges, which, save in the very worst of droughts, yields reasonable rains. That it was always fertile is evidenced in its coal measures, the tamped-down, petrified relics of carboniferous jungles; these, coupled with a rich, extremely wide alluvial plain, have ensured that the Hunter has always flourished under European settlement, its agricultural and pastoral activities of equal importance to its industry. But the Hunter isn't navigable except at its mouth, and even there, huge sand bars foul it.

Once, on a train 1600 kilometres from its outflow into the Gulf of Mexico, I crossed the Mississippi River at its confluence with the Missouri River. *It was 3 kilometres wide.* You could have fitted all of Australia's inland rivers into it, and had but a puddle on its bed.

WATER IS LIFE

WATER IS LIFE

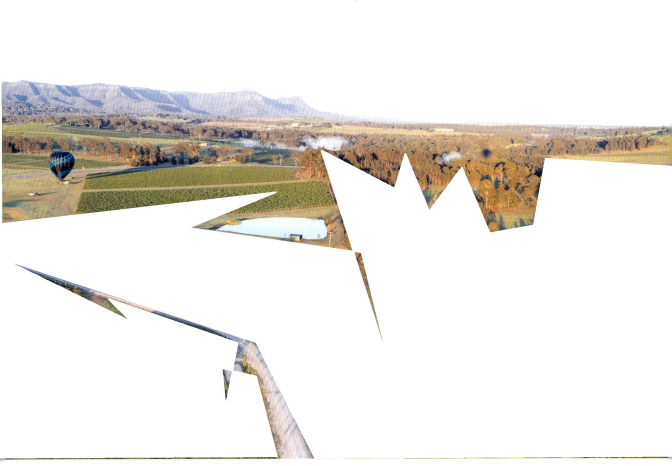

The central Hunter Valley has hundreds of irrigated vineyards and is recognised in Australia and abroad as a major producer of high-quality varietal red and white wines.

WATER IS LIFE

WATER IS LIFE

Glenbawn Dam, near the upper Hunter Valley township of Scone, is a prime source of irrigation water for horticulture. The district is known for its beef, grain and thoroughbreds.

The Mississippi River at its confluence with the Missouri River, some 1600 kilometres from its outflow into the Gulf of Mexico.

I mention this because I had what I term an 'Australia moment' – the dawning of a truth about Australia not seen before. And it made me understand the manifest differences that the land of one's birth and upbringing must make on one's perception of the world. I looked at the Mississippi River and harkened back to living on the Darling River as a kid, and of throwing a stone across it onto the far bank, and of the worry on the men's faces when the river dwindled to a few muddy pools along a satiny, cracked bed. There, in that train rumbling across a seemingly endless bridge over what appeared a ship-dotted sea, I realised that America had greatness literally thrust upon it. Five vast inland seas and thousands upon thousands of kilometres of interlinked, navigable rivers fusing the continent's heartlands into a natural endowment of staggering scope. The USA *is* a land of milk and honey, a gargantuan bread-basket, a place of bounteous water; even 'minor' rivers like the eastern Hudson and Connecticut are hundreds of kilometres long, wide enough to carry little icebergs in a thaw.

At that moment I saw Australia in perspective, the good with the bad, the tangible against the dream. An unforgiving and unkind land, surely all the more worth winning for that. I saw the reason for two such divergent histories: having so much, the USA could throw off the British yoke early, and with impunity; whereas Australia, having by comparison so little, was better off sticking with the British Empire. A choice which doesn't render Australia less individual or Australians less than unique. But what it screams loud and clear is obvious: *water is life*!

WATER IS LIFE

Michael Crouch genuinely believes that Australia is the best place on Earth, and its people the best people. This intense, positive, all-embracing patriotism springs from the very core of him, and he has dedicated his life to it, no matter which way around you turn the multifaceted epics of his life's journey. His products must be the best, Australian products are the best, therefore his products will be made in Australia; no poet thrills him like the Australian bush poets, for they lyrically express what he feels for Australia; his life-style is 'old' Australian in that he'd rather savour the smell of gum trees than car exhausts; his philanthropies and his interests are Australian.

Fairly early in 1996, Michael received a phone call from John Howard, the Prime Minister of Australia, that was to lead to a long-term, time-consuming load of voluntary work of a kind that gets little

> **Michael Crouch genuinely believes that Australia is the best place on Earth, and its people the best people.**

Michael Crouch at the APEC Leaders' Meeting held in November 2006 in Hanoi, pictured with George W. Bush, President of the United States of America 2001–09.

Michael Crouch with the Honourable John Howard OM AC, Prime Minister of Australia 1996–2007, who appointed Michael in 1996 to be one of his three representatives to the ABAC.

public thanks and few laurels. Yet of everything Michael Crouch has achieved, his appointment to APEC as one of Australia's three delegates remains his proudest accomplishment. Other Australian delegates would come and go, but Michael held his position for eleven years, from 1996 to 2007.

I speak of APEC (the Asia-Pacific Economic Cooperation) and its intrinsic function, ABAC (the APEC Business Advisory Council) at this point in Michael's story because it best illustrates the consuming passion of his patriotism. What CEO in his right senses would take on an unpaid job as his nation's representative for five lots of talks per year, each talk scattered somewhere around the Pacific's mighty bowl? But Michael managed for eleven years; three of his early colleagues, Imelda Roche AO, Malcolm Kinnaird AC and David Murray AO each served for approximately five years, and there were other selfless patriots too, such as Mark Johnson AO, who served for more than eleven years.

APEC was a sort-of-answer to NATO, though its objectives and achievements were less headline-snatching than Big Atlantic Brother. It contained twenty-one nations, including the USA, China, and Japan, so it had plenty of muscle and clout. What it set out to do in ABAC was to remove the bureaucratic barriers between a nation's political heads and its business heads, and though progress was slow, it was steady. Michael's chief contribution lay in raising the consciousness of individual nations to standards, which are the criteria governing the importation of foreign goods. At first, he got blank stares, but over the years he succeeded in making less standards-conscious nations realise that in order

WATER IS LIFE

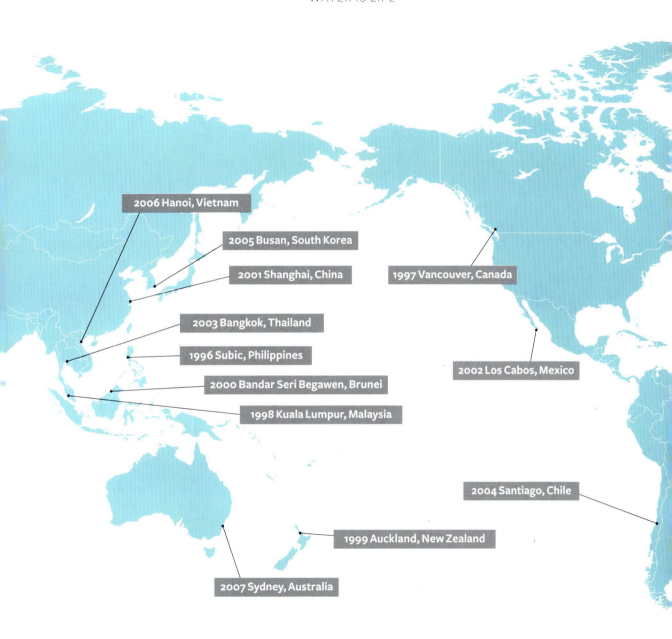

APEC meetings around the Pacific Rim during the period of Michael's APEC appointment.

to have a healthy export trade to a specific country, they must adhere to its standards. The USA, Australia and Japan, for example, maintain stringent standards.

Four of the five annual meetings were directly between ABAC members, while the last was a private meeting between a nation's political leaders and its APEC delegates. All up, his APEC duties consumed about six weeks annually of Michael's time. The location of one particular meeting might be incongruous and awkward – but it saved the Premier of the host nation from standing too long at a time on his feet – he was old and feeble, but dare not let his people know that!

What APEC did for Michael personally was to imbue him with a huge sense of pride in Australia; he sat, one of many delegates, only a handful of whom were Caucasian, watching and listening in awe. Aside from honing his patriotism, APEC gave Michael a profound education in the world as it is at the dawn of the Third Millennium. Looking more and more at dwindling global water. Water is life!

WATER IS LIFE

In 1788, ensconced at Sydney Cove, the First Fleeters took over the Tank Stream, good water. It's well known that the First Fleeters almost starved to death because no other ships came for two and a half years. What isn't nearly as well known is that a supply ship, the *Guardian*, stuffed with goodies, was distantly following in the wake of the First Fleet. Her decks choked with horses and cattle, the *Guardian* left Cape Town and began to search for her eastings, a tortuous business. Having found hints of them only, the ship inched down into the Roaring Forties. A thousand miles out, still floundering in cat's-paw breezes, Captain Riou started to panic; the livestock were drinking up his fresh water at an alarming rate. Then – manna from heaven! The Captain spied a gigantic iceberg on the far horizon, and had a brilliant notion. Sailing in much closer to it, he sent boats loaded with empty water tuns to land on the ice, their crews under orders to chip ice off the berg and fill the tuns. While the sailors set to with a will, Captain Riou retired below to enjoy a boozy lunch, apparently unaware of the fact that only one-ninth of an iceberg sits above the surface of the sea. Colliding with a submerged spur of ice, the *Guardian* stove in her round-tucked stern and mangled her rudder. After a harrowing ordeal, a handful of the survivors reached the Cape to transmit the ship's fate to the Admiralty. In the meantime, the hapless First Fleeters watched in vain for the *Guardian*'s sails. The truth of the matter was that these routine Royal Navy captains were ill-suited for adventuring into unknown places; they had not the training for it.

The story of the *Guardian* was an omen pointing to the critical role water would play in the history of European Australia. No land replete with water or waterways, but there were some sources of water, and many ingenious minds to deal with them.

Some inland regions had a shallow water table. If a hole were bored to tap into its sweet water, that water was pumped up using wind power – a tall derrick supporting blades moored to a platform that turned according to wind direction. When drilling equipment became available to bore holes some thousands of metres deep, the vast artesian water basin was tapped too. Roaring and near-boiling, the water erupted under huge pressure, reeked of sulphur, and was not fit for humans to drink. But it was a godsend nonetheless.

After which came the heyday of the big dams, a boon to rural people in many ways. Combined with new drought-resistant crop

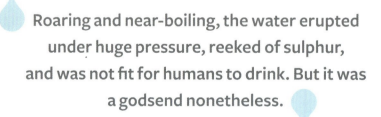

> Roaring and near-boiling, the water erupted under huge pressure, reeked of sulphur, and was not fit for humans to drink. But it was a godsend nonetheless.

WATER IS LIFE

Severe droughts occur in many regions of Australia, rendering croplands and pastures useless for several years.

WATER IS LIFE

Windmills pumping water from aquifer basins for stock watering and crop irrigation remain a commonplace sight in 'outback Australia' but aquifers were once the only water resource available to a majority of farmers and graziers.

strains, irrigation saw Australia in better case to contribute food to a mushrooming world population. Unfortunately, non-food crops became more profitable for growers, which leads to the as yet unanswered question: how much control over dam water should governmental agencies have? Can the growing of cotton be deemed a wastage of irrigation water?

Some men had imperishable dreams. One such was to divert the short little rivers of the snow country into the Murray-Darling river system, augmented by big dams on tributaries like the Namoi and Murrumbidgee. The cost was astronomical, but it was done. Today certain environmentalists consider the Snowy Scheme an ecological crime, but it was a product of its times and undertaken in good faith. It provided many thousands of jobs for at least two decades, and it enabled the Murray-Darling-Murrumbidgee to become a bread-basket. Like the Aswan Scheme in Egypt, what promised to be an improvement betrayed an increasing number of false premises resulting in unlooked-for disasters, but surely the human race can learn from previous mistakes in a positive way?

The more ambitious Australian sister scheme was never tackled. That had been to divert the short, deep, year-round rivers of coastal North Queensland through the Great Divide to pour water into the Cooper, the Diamantina, the Barcoo and other mostly dry rivers of western Queensland, where nigh limitless black soil plains need but one thing to be a bread-basket – water. However, its cost was beyond the financial resources of the nation. Though it's tempting to wonder how the accomplishment of this scheme would have altered the water profile of the Australian interior. Would the negative aspects outweigh the positive? Unless some scholar is doing a massive computer simulation of it taking all the pluses and the minuses of the Snowy into account, we'll never know.

In the Third Millennium the talk is all about greenhouse gases, climate change, acidified oceans, and melting everythings. Among the direct results of over-population are nightmares like exploitation of the ignorant, industrial and vehicular pollution, and acid rain. Some few problems are peculiar to Australia, all intimately connected to our great lack – water storage.

WATER IS LIFE

As a boy and a youth, Michael kept, not a diary, but a collection of aphorisms, poems, foreign language phrases and pithy truths chosen for one overriding reason: they contain inspirational ideas about goodness, the fruits of hard work, the ability of the human mind to soar above its animal origins. The foreign phrases all hold a kind of verbal music (he jokes, for instance, about onomatopoeia, which a word owns if it sounds like its meaning, as in 'buzz' or 'whoosh!'). His favourite Gospel is Matthew; his favourite poets Siegfried Sassoon, Alfred Tennyson and Horace; his favourite philosopher is Shakespeare. The 'jottings' – his name for the collection – are not what one would expect from a toughly enduring businessman.

But then, Michael Crouch doesn't fit into any preconceived mould, of businessman or other.

His philanthropic efforts are anchored in Australia and things Australian. Typically, his approach to his philanthropy is hands-on. To Michael, it's not just a matter of writing a cheque and having his name chipped out in gilt letters on a marble slab as a Big Donor. He believes education and research lie at the heart of Australia's future. At the University of New South Wales he commissioned Australia's first Chair of Innovation in its School of Business, of whose Advisory Board he had been a member for ten years. Later he caused the University of New South Wales to establish its Innovation Centre, and at the University of Sydney's Brain and Mind Research Institute, he caused the first Chair of Depression and the first Chair of Children's Mental Health to be established.

Under his energetic aegis the Royal Flying Doctor Service's drive for two more planes came rapidly to fruition; that out of many deserving projects Michael chose to ally himself with the Royal Flying Doctors is rooted in his pride that the people of the Australian outback have a flying doctor to call. The same is true of the Duke of Edinburgh Awards: it is Australian youth in particular who benefit from his efforts.

Though you will have to look hard to find Michael Crouch in the group photographs occasionally taken to commemorate charitable endeavours; if you scan intently, you will find him in the back row's most insignificant spot.

Michael's musicality has already been mentioned, in that he played the piano to give himself pleasure as a child – rare in a boy, even one with a history of chronic illness. But it does mean that Michael has an intense love of music, and he has helped to further the careers and ambitions of young musicians of great musical promise.

When an extremely young Australian singer-composer, Gavin Lockley, was drawn to his attention, Michael encouraged him to write a Symphony of Australia. This work was performed at the Sydney Opera House to celebrate the eightieth birthday of the Royal Flying Doctor Service, and was enthusiastically received.

A concert to celebrate the 85th Birthday of
The Royal Flying Doctor Service of Australia

Staged in partnership with Sydney Festival

My Country Australia

Celebrate what it means to be Australian!
Hear our favourite poets come alive in song…
Dorothea Mackellar, Banjo Paterson, CJ Dennis, Henry Lawson

Featuring Jon English, Renae Martin, Barry Ryan, Glenn Cunningham, Darryl Lovegrove, Jennifer Murphy, Michael Halliwell, Tina Harris

Plus 60 Piece Orchestra and a Choir of 200 Voices

Hosted by
Stuart Maunder
Conducted by
Vladimir Fanshil
Music Composed by
Gavin Lockley

31 October 8pm
Concert Hall, Sydney Opera House

Bookings
sydneyoperahouse.com | 02 9250 7777

PRINCIPAL SPONSOR
Zip
Instant Boiling Water

SUPPORTING SPONSOR

Gavin Lockley then went on to set the famous Dorothea MacKellar poem 'My Country' to music. Michael hopes to see it become as well known as our national anthem. Simply, he feels it is more emotionally descriptive than 'Advance Australia Fair'.

Michael and Gavin share an all-important aim: to build a great treasure-house of Australian culture through music and song.

Gavin read my novel *Morgan's Run* and asked me for the right to set it to music as something between an opera and a musical. I consented, provided I could write the libretto. Michael came aboard immediately, doing whatever he could to assist Gavin. After much hard work and the passing of several years, *Morgan's Run* the opera-musical had a workshop production in a small town near Sydney. The audience greeted it with amazing enthusiasm, warts-and-all production though it was. At the time of writing, Gavin is putting it into oratorio form.

There are more charities under Michael's capacious wing; how many there might eventually be is in the lap of the gods. But if some Australian person, activity or thing comes to Michael's notice, he'll be off down that path throwing all of his heart into every step, no matter how stony or slippery the way.

WATER IS LIFE

Shanny Crouch, Michael Crouch and the Sydney Opera House.

WATER IS LIFE

WATER IS LIFE

The cast of *Morgan's Run: The Musical* at its world premiere, in May 2011 at Springwood, New South Wales. The story of one man's epic First Fleet voyage from England to Australia and the founding of the colony, *Morgan's Run: The Musical* was written by Colleen McCullough, with music composed by Gavin Lockley, and performed by members of the Blue Mountains Musical Society. Michael Crouch acted as its premiere Executive Producer.

WATER IS LIFE

What about the future of water in Australia? Some allusions to that future have been made in passing, but here is the place to dwell on it, as Michael Crouch does.

The gift of foresight in human nature is rare; not realising that they do so, people live from moment to moment in the comfort of thinking that conditions for future generations will be as easy or even easier than their own. That the conditions of the past were always worse. But is that comfort mistaken?

I will use a double standard, comparing the modern water profiles of the USA and of Australia.

In 1950, the population of the USA stood at 150 million. By 2000, that population had more than doubled. Despite its plethora of big cities, the USA is a nation of towns, farms and pastures; its heartlands are quite densely populated. But even the lushest bread-basket river systems have a limit when more and more food and plant products have to be grown with the assistance of irrigation. The waters of the Colorado and some other western rivers, dammed and tamed, are sent as irrigation to the vast food cultivation areas of California. Ever-increasing demand for water has reduced the Colorado's spate so much that it no longer reaches the sea. The same kind of thing is happening in other areas too. Not as the result of climate change, but of sheer demand for irrigation water. There are more than twice as many mouths to feed as there were fifty years ago.

In 1950, the population of Australia stood at 8 million. By 2000, it had more than doubled to 19 million. In 2012, it has grown to over 22 million. Immigration has been a very big factor. During those fifty years, 1950 to 2000, a precarious bread-basket river system was restructured to supply large-scale irrigation, but the heartlands are incapable of supporting even the most modest of human populations. Eighty-five per cent of Australia's people live in sprawling coastal urban communities unattached to bread-basket systems.

Urban water comes from the damming of every possible local river, and in the past urban excremental waste was often pumped untreated into the ocean. Why untreated sewage? Because urban population growth had far outstripped the erection of the sewage treatment works needed.

Rural folk have dwindled so drastically that their political arm, the Country Party, changed its name to the National Party. Only urban centres possess vote reservoirs large enough to matter politically; there are more members of the parliaments seated on or around the Cumberland Plain than seated in the rest of New South Wales. City people matter. Country people do not. And that leads to a national populace grossly ignorant about the intrinsic qualities and functions of water. Its politicians, whose foresight extends only to getting re-elected, in the main share this ignorance.

Australian topography shapes its perennial river systems: short rivers stretch from mountains to the sea on the northern, eastern and southern coasts, while the massive Murray-Darling Basin stretches from Queensland in the north to the Murray River's Southern Ocean outfall in South Australia. Non-perennial rivers in northern monsoonal regions and in dry inland areas (such as those draining into Lake Eyre) are not included here as frequently they remain dry for years.

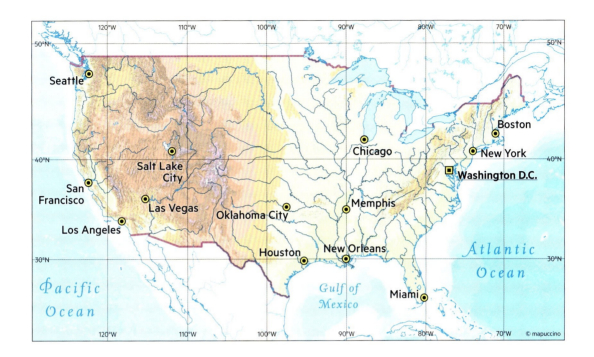

A similar topographic map of continental USA indicates how its river systems drain a remarkable number of reliable mountain water catchments to eastern, western and southern coasts, while some of its largest river systems also provide immense navigable waterways traversing much of the nation from west to east and from north to south.

WATER IS LIFE

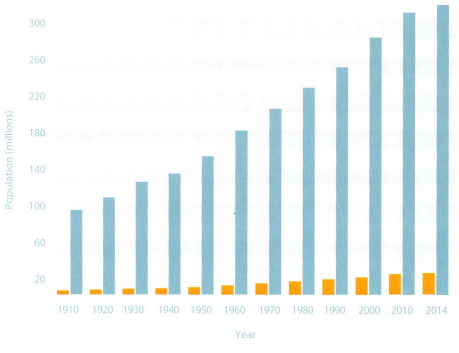

Population and land area of Australia and the USA

Australia land area: 7 692 024 square kilometres

USA land area: 9 147 593 square kilometres

A comparison of the area and population statistics of Australia and continental USA shows that while the area of the USA is 19 per cent greater than that of Australia, its population has grown to be 12.5 times greater and continues to expand at a rapid rate.

How does an urban Australian view Australia?

As a wonderful place of democratic freedom having a sunny climate, a glorious coastline, every kind of food and sport and entertainment any reasonable person could want, perhaps a job, otherwise good welfare, and, in all cases, water gushing out of taps into basins, buckets, bowls, baths, shower-stalls and toilets, not to mention onto lawns and flower beds.

How are they informed about water? The news media give them statistics about droughts, including footage of dry rivers and dead animals, but it's all second-hand, not directly experienced. Water rationing means no automatic sprinklers in the garden, and Government spends some of its money advertising to conserve water rather than waste it. Yet it isn't the plight of the bush farmer that really worries the urbanite: that is his/her own plight as the water level in the local dams supplying his/her own city keeps dropping.

Australia sits like an unbalanced dish, its internal expanses sparsely settled in some parts, but more of it utterly bare; two-thirds of its people living on the narrow coastal strip between Geelong–Melbourne and Cairns, while the vast majority of the other one-third live in two separate clusters – coastal South Australia and the south-western corner of Western Australia. The dish is in permanent danger of tipping to rest on its eastern side, where the weight of the populace is. And did that happen, no inland seas would empty away, because there are none, and the rivers are trickles. In rural areas the water comes to Earth and has a chance of staying there, but in coastal urban areas much of the water is lost forever into the ocean.

In this rests Michael's chief frustration. He feels, and rightly so, that Australia's political and bureaucratic arms of government do not take the Australian water situation seriously enough: that some lobbies have disproportionately loud voices, and that the ordinary Australian no voice at all when it comes to water. To sell a nation's birthright is always the province of its two governing arms, and may or may not be wise, or necessary, or whatever. But water is life, and Michael *knows* that Australia must do far more, spend far more, on ensuring a future supply of water than Australia has done or is prepared at the moment to do. For the working man to enjoy the best standard of living on the planet is laudable, but how much thought is given to the one thing that can guarantee that – water?

The theme of Martin Luther King Junior's greatest speech is 'I have a dream!' Words that echoed around the world to reach far vaster audiences than his own African-Americans. *Everybody* heard him. Michael Crouch's fiercest and most driving dream is that his country, Australia, solves its age-old problem, shortage of water.

Who can say? Australia has already built a small number of plants to desalinate sea water, with more under construction. And in the meantime, storing far more water across our landscape and even piping water from the north into more arid parts of the continent would be a start. Nothing is impossible. Ask Michael. He knows.

WATER IS LIFE

In concentrating upon the story of water, I have tried to illustrate the life of one man, Michael Crouch. An important man, a man who has made his mark in so many different and admirable ways. Long after he has sloughed off this mortal coil, his influence will continue to be felt as a champion of innovation, in addition to his advocacy in the fields of water distribution to human habitation, the techniques of water conservation, the heightened awareness of some Pacific nations to regional politico-business conferences, university and research endowments, his contributions to Australian music and the arts, and his ardent patriotism. And that is just the tip of the iceberg.

Michael and Shanny Crouch in 2012.

Major Australian Water Projects

Snowy Mountains Hydro-electric Scheme

LOCATION	South-eastern New South Wales
STARTED	1949
COMPLETED	1974

COMPRISES
- 16 major dams
- 7 power stations
- 1 pumping station and 1 pump storage facility
- 225 kilometres of tunnels, pipelines, aqueducts

FUNCTIONS
- Diverts Snowy River flow from south to west
- Releases into the Murrumbidgee River
- Releases into the River Murray
- Supplies power to the National Electricity Grid

Snowy Mountains Hydro-electric Scheme

Upper reaches of the Snowy River are fed by an annual snow melt.

ABOVE The Murray 1 Hydro-electric Power Station, located near the town of Khancoban, in the Snowy Mountains. It is a gravity-fed hydro-electric water station receiving water from the Geehi Reservoir, on the Geehi River, above the power station and discharging it into the Murray 2 Pondage beneath.

LEFT The machine mall floor of the Murray 1 Hydro-electric Power Station, initially commissioned in 1967.

Waters from the Snowy, Murrumbidgee, Darling and Murray rivers reach the sea in South Australia.

Murrumbidgee Irrigation Area

LOCATION	South-western New South Wales
FOUNDED	1912
	Burrinjuck Dam construction commenced 1906
	Canal construction continued until 1920
	Expanded to adjacent districts until 1938
COMPRISES	• 660 000 hectares in total
	• 170 000 hectares irrigated
	• 3200 landholdings
	• 2500 owner irrigators
	• 3500 kilometres of supply channels
	• 1600 kilometres of drainage channels
FUNCTIONS	• Diverts river water for food and fibre production
	• Global export of fruits, wines, grains, vegetables, cotton, wool

The Murrumbidgee Irrigation Area is a major source of fruit crops.

Burrinjuck Dam, created to feed the Murrumbidgee Irrigation Area.

The Murrumbidgee Irrigation Area has over 3200 irrigated landholdings.

Ord River Scheme

LOCATION	North-eastern Western Australia
STARTED	1959
COMPLETED	Kununurra Dam, 1963
	Lake Argyle, 1972
COMPRISES	• 2 major dams
	• 1 power station
FUNCTIONS	• Captures Ord River monsoonal flows
	• Irrigates 12 500 hectares of farmed area
	• Potential to irrigate 45 000 hectares

The Ord River Scheme irrigates about 12 500 hectares and has the capacity to irrigate over 45 000 hectares.

The earth- and rock-filled Ord River Dam, in Western Australia's Carr Boyd Ranges, forms a major storage reservoir, Lake Argyle. It is one of the world's largest human-made water bodies, with a capacity of 10 763 gigalitres (about twenty-one times the estimated volume of Sydney Harbour).

Cubbie Station

LOCATION   Southern Queensland

COMPRISES
- 96 000 hectares in total
- 22 000 hectares developed
- 11 000 hectares under development

FUNCTIONS
- Converted to cotton production 1983
- Largest irrigated property in Southern Hemisphere

The Culgoa River, near Dirranbandi, Queensland, is the prime source of water for irrigation at Cubbie Station.

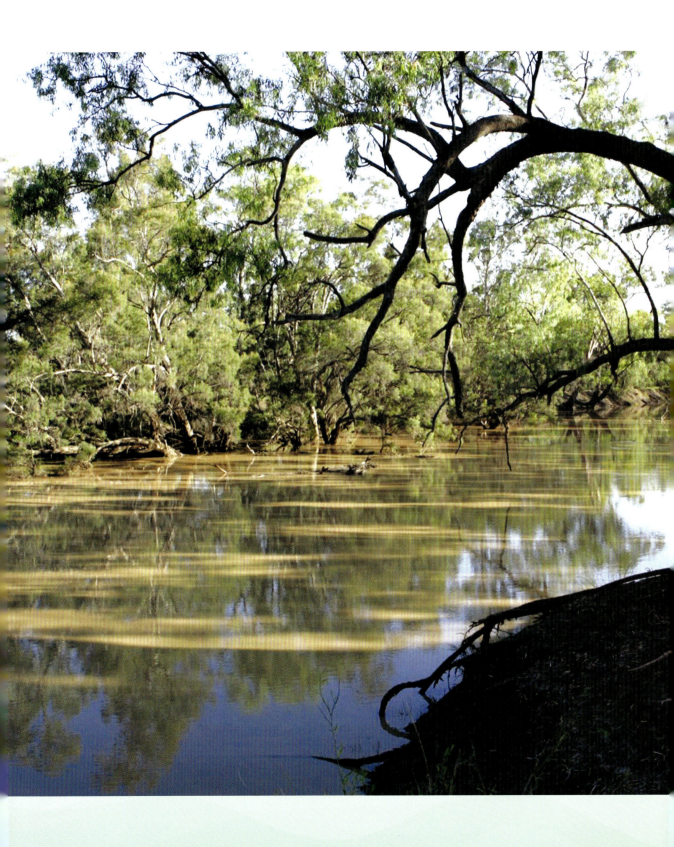

Irrigated fields of cotton, stretching as far as the eye can see.

Cotton harvesting at Cubbie Station.

Bradfield Scheme

LOCATION North-eastern Queensland
DEVISED BY Dr John Bradfield, 1938
STARTED Never

FUNCTIONS
- Diverts Tully, Herbert and Burdekin river flows
- Feeds Thomson River flowing west
- Provides water to irrigate 1000 million hectares
- Eventually feeds into Lake Eyre

The Tully River would be diverted under the Bradfield Scheme.

Water spills over the Burdekin Dam wall at Ravenswood, south-west of the northern Queensland towns of Ayr and Home Hill, in December 2010.

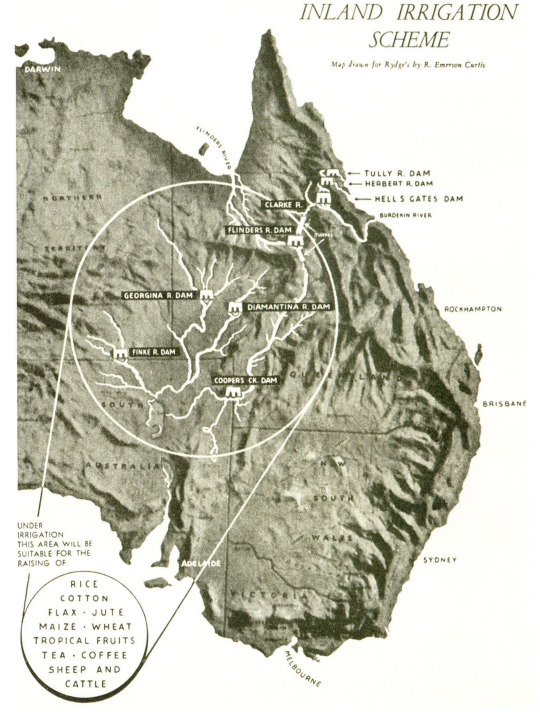

Picture Credits

Endpapers Getty Images
Page iii Getty Images
Page iv Zip Corporate Archive
Page 2 Getty Images
Pages 13, 14, 17, 18, 19 Zip Corporate Archive
Pages 20–21, 24, 26–27 Getty Images
Pages 30–31 Science Photo Library
Pages 36–37, 38, 39, 40 Zip Corporate Archive
Pages 42, 44 Australian Government Bureau of Meteorology
Pages 45, 47, 48 Getty Images
Pages 53, 54–55, 56, 57, 62 Zip Corporate Archive
Pages 69, 70–71, 72 Crouch Family Collection
Pages 74–75 Matt Lauder Gallery
Pages 76–77 Peter Bellingham Photography
Page 78 Getty Images
Page 81 Zip Corporate Archive
Page 83 Kim Gregory

Pages 87, 88 Getty Images
Page 93, 95 Zip Corporate Archive
Pages 96–97 Gavin Lockley Collection
Pages 100, 101 Gavin James, Mapuccino
Page 102 NewSouth Publishing
Page 105 Crouch Family Collection
Page 109 Getty Images
Page 110 Brad Chilby Photography
Page 111 Getty Images
Pages 113, 114, 115 Murrumbidgee Irrigation Limited
Pages 116–17, 118–19 West Australian Newspapers
Pages 120–21 Getty Images
Pages 122, 123 Cubbie Station Collection
Pages 124–25 David Foster/Alamy
Page 126 Cameron Laird Photography
Page 127 *Walkabout Magazine*, January 1947/State Library of NSW

Index

Page references in **bold** indicate an illustration.

ABAC (APEC Business Advisory Council) 81, 82, 84
Africa 58
Albert Edward, Prince of Wales 5–6
America 79, 98, **101**, **102**
ancient Romans 2, 3–4, 59
animalcules 5
animals 42–43
ANZ Bank 10
APEC (Asia-Pacific Economic Cooperation) 52, 82–84, **83**
APEC Business Advisory Council 81, 82, 84
appliance manufacturers 16
aquifers 88
Argyle, Lake, WA **120–21**
artesian basins 86
Asia-Pacific Economic Cooperation 52, 82–84, **83**
atoms 22
Australia
    population distribution 98, 99, **102**, 103
    rainfall 44, 45, 46, 48
    rivers **100**
    topography **100**
    water profiles 46, 60, 79, 86, 98, 104
Australian Design Award **40**
Australian Labor Party 33
Australian manufacturing 16
Australian water projects 108–31

bacteria 5
basins, water 88
bathrooms 8

bathroom water heaters 8, 10, 12, 15
bathtubs 7
Bigger, George 7–9, 67
Blue Mountains Musical Society **96–97**
boiling water 6, 34
bore water 86
Boulton, Matthew 34

Bradfield Scheme, Qld **124–27**
bread-basket river systems 43, 46, 59, 73, 89, 98, 99
bread, origins 58
British Empire 79
Burdekin Dam, Qld **126**
Burrinjuck Dam, NSW **114**
Bush, George W. **81**
business standards 84

carbon 29
Centigrade scale 23
central Hunter Valley, NSW **74–75**
Chair of Children's Mental Health, Brain and Mind Research Institute, University of Sydney 92
Chair of Depression, Brain and Mind Research Institute, University of Sydney 92
Chair of Innovation, School of Business, University of New South Wales 91–92
chilled filtered sparkling water units **54–55**, **56–57**
chilled filtered water units 52, **53**, **54–55**

chip heaters 8
cholera 4, 5
Colorado River, USA 98–99
Comet McNaught **24**
comets 23–28
compounds, chemical 22, 23
condensers 34, 35, **38**
conservation, water 60–61, 103
conservative government, Australian 33
coppers 7
cotton 89, **123**
Country Party of Australia 99
Cranbrook School, NSW 65
Creation Moment 22
Crimean War 4
crops, non-food 89
Crouch, Charlotte 70, **72**
Crouch, George 70, **72**
Crouch, Michael 9, 11, 15, **17**, 18, 52, 63–72, **69**, **81**, **95**, 104, **105**
    ideals 50–51, 71
    Morgan's Run: The Musical (McCullough/Lockley) 94, **96–97**
    patriotism 80, 82, 84
    philanthropy 91–94
Crouch, Sarah 70, **72**
Crouch, Shanny 68, **69**, 70, 72, **95**, **105**
Cubbie Station, Qld **120–23**
Culgoa River, NSW/Qld **120–21**
Cumberland Plain, NSW 99

dams 86, 89
Darling River, NSW 79
desalination plants 104
deserts 46
'dirty snowball' hypothesis 25, 28
diseases 4, 5
disinfectants 5
diversification 52
diversion, water 89

drinking–excretion cycle 3–5
drinking water 3–6
droughts 87
Duke of Edinburgh Awards 92

Earth 29, **30**
education 6, 103
Egypt 58
electricity 12, 15
electrification 12
electrons 22
elements, chemical 22, 29
emotions 3
energy 42, 60–62
engines, steam 34
enteric fevers 4–5, 60
environmental issues 60–62
Euphrates River, Middle East 58
explorers, European 60
export trade 84

faucets (taps) 51
filters, water 52
First Fleeters 85
First World War 5
food poisoning 60
forests 46, **47**

gas giants 25
germs 5
Glenbawn Dam, NSW **76–77**
Gondwanaland 41, **42**, 43
government, Australian 33, 103, 104
grasses, edible 58, 59
Great Comet of 2007 **24**
Guardian, HMS 85–86

Halley, Edmund 25
Halley's Comet **26–27**
heaters, water *see* water heaters

INDEX

Hoover Award for Marketing 15
hot water 7, 60
hot water storage systems **14**
hot water systems 15–16, 34
hot water urns 7–8
Howard, John 52, 80, **81**
human beings, first 58–59
human waste disposal 59, 99
Hunter River, NSW 73
Hunter Valley, NSW **70–71**, 72, **74–75**, **76–77**
hydrogen 22
hygiene 3–6, 59–60

icebergs 85
ice planets 25
inflation 33
Innovation Centre, University of New South Wales 92
instant boiling water heaters 34, 35, **36–37**, **38**, **40**, 52, **53**, **54–55**
international distribution 35
irrigation 88, 89, 98–99

Johnson, Mark 82
jungles 43

Keating, Paul 51
kettles 7
King, Martin Luther, Junior 104
Kinnaird, Malcolm 82
kitchen water heaters **14**, 49, 51
Kuiper Belt 25, 28, **30–31**

Labor Party, Australian 33
Laurasia 41, **42**, 43
Leeuwenhoek, Anton van 5
life 29, 32, 42, 43, 46, 59
Lister, Joseph 5
Lockley, Gavin 92, 94, 95, 97
    Morgan's Run: The Musical 94, **96–97**
    Symphony of Australia 92, 95
Lodge Cranbrook 67
London, United Kingdom 5
low pressure water heaters 10

MacKellar, Dorothea, 'My Country' 94
mammals 43
manufacturers, appliance 16
Marrickville, NSW 10
Mars 29, **30**
McCullough, Colleen 94, 97
    Morgan's Run 94
    Morgan's Run: The Musical 94, **96–97**
Mediterranean basin 41
Mercury 29, **30**
Mesopotamia 58
microbes 5, 6
Mississippi River, USA 73, **78**, 79
Missouri River, USA 73, **78**
molecules 22, 23, 29, 42
monsoons 46
Morgan's Run (McCullough) 94
Morgan's Run: The Musical (McCullough/Lockley) 94, **96–97**
Murray 1 Hydro-electric Power Station **110**
Murray-Darling Basin 89, 100
Murray, David 82
Murrumbidgee Irrigation Area, NSW **112–15**
'My Country' (MacKellar) 94

National Party of Australia 99
Nemausus (Nimes), France 2
New South Wales, rural 12
New Zealand 7
Nightingale, Florence 4
Nile River, Africa 58
Nimes aqueduct, France **2**
Nimes, France 2

Northern Hemisphere 41

Oort Cloud 25, 28
Ord River Dam, WA **118–19**
Ord River Scheme, WA **116–19**
oxygen 23

Pangaea 41, **42**
patents 8, 10, 15, 33, 34
Periodic Table of the Elements 23
planets 25, 29, **30–31**
plants 42–43
Pont du Gard, France **2**
population distribution, Australian 98, 99, **102**, 103
potable water 3–6
profiles, Australian water 46, 60, 79, 86, 98, 104
projects, Australian water 108–31
protons 22
proto-planetary disc 25, 29

rain 29, 46
rainfall, Australian 44, 45, 46, 48
rationing, water 103
recession 51
research and development 33, 50
Riou, Edward 85
rivers 46, **100**, 111
Roche, Imelda 82
rocky planets 25, 29
Romans, ancient 2, 3–4, 59
Royal Flying Doctor Service 92
rural New South Wales 12

scientists 3
shower water heaters 10, 12, 15
society, origins 59
Snowy Mountains Hydro-electric Scheme, NSW 89, **108–11**

Snowy River, south-eastern Australia 108, **109**
solar system **30–31**
Southern Hemisphere 41
sparkling water units **54–55, 56–57**
standards, business 84
steam 34, 38
steam condensers **38**
steam engines 34
sterilising units 6
sterility 6
storage, water **14**, 90
Sun 25, 28, 42
surf beaches **48**
Sydney, NSW 12, **20–21**
Sydney Stock Exchange 16
Symphony of Australia (Lockley) 92, 95

tables, water 86
Tank Stream, NSW 85
taps 51
teamwork 32
tectonic plates 42
TEDx 'Ideas worth spreading' conference 56
temperature, measurement 23
Tigris River, Middle East 58
topography, Australian **100**
Tully River, Qld **124–25**
twin tanks 35
typhoid 4, 5–6
typhus 4

United States of America 79, 98, **101**, **102**
universe 22
University of New South Wales 91–92
University of Sydney, NSW 92
upper Hunter Valley, NSW **70–71**, 72, **76–77**
urban Australians 103
urban housing 12

INDEX

urban water 99
urns, hot water 7–8
USA (United States of America) 79, 98, **101, 102**

valves (taps) 51
Venus 29, **30**

wage increases 33
washing 60
waste disposal, human 59, 99
water 3, 29, 46, 59, 99, 103
    chemical formula 23
    origins 25, 28
water basins 88
water condensers 35
water conservation 60–61, 103
water diversion 89
water heaters
    bathroom 8, 10, 12, 15
    chip 8
    installation 49
    instant boiling 34, 35, **36–37, 38, 40**, 52, **53, 54–55**
    kitchen **14**, 49, 51
    low pressure 10
    shower 10, 12, 15
water, potable 3–6
water profiles, Australian 46, 60, 79, 86, 98, 104
water projects, Australian 108–31
water rationing 103
water storage 90
water storage systems **14**

water tables 86
water, well 4
Watt, James 34
Waverley Station, NSW **70–71**
well water 4
Whitlam, Gough 33
windmills **88**
wind power 86

Yarra Ranges 46, **47**

Zenith Pty Ltd 68
Zip Heaters (Aust.) *see* Zip Industries Group
Zip Heaters (UK) 68
Zip Hydroboil **39**, 49
Zip HydroTap 51–56, **53, 54–55, 56–57, 62**
Zip Industries Group (previously Zip Heaters (Aust.)) 8–17, 32, 35, 49, 50–51, 60–62, 67, 68
    Australian manufacture 51, 71, 80
    awards 15, **40**, 49, 52
    design 15, 35, 49
    executives 50, 70–71
    expansions 10–11, 50, 68
    factory **13, 17**, 50
    improvements 15, 50, 52
    international distribution 35, 68
    marketing 10, 11, 33, 61
    premises **13, 17**, 50
    profits 11, 16, 32–33
    research and development 33, 50
    staff 32, 51
Zip Miniboil 35, **36–37, 38**, 49